Investigating Human Can[c]
Computational Intelligence Techniques

Investigating Human Cancer with Computational Intelligence Techniques

Alfredo Vellido
Paulo J.G. Lisboa

KES International Recent Research Results Series

Future Technology Press
United Kingdom

©2009, Future Technology Press

Investigating Human Cancer with Computational Intelligence Techniques

Volume Editor:
    Alfredo Vellido
    Soft Computing Research Group, Technical University of Catalonia,
    C. Jordi Girona, 1-3. 08034, Barcelona, Spain.
    Email: avellido@lsi.upc.edu

    Paulo J. G. Lisboa
    School of Computing and Mathematical Sciences,
    Liverpool John Moores University
    Byrom St. L3 3AF, Liverpool, United Kingdom
    Email: P.J.Lisboa@ljmu.ac.uk

KES International Recent Research Results Series

Series Editors:
    Robert J. Howlett
    University of Brighton, School of Environment and Technology,
    Moulsecoomb, Brighton, BN2 4GJ, United Kingdom
    Email: rjhowlett@kesinternational.org

    Lakhmi C. Jain
    University of South Australia, School of Electrical and Information
    Engineering, Mawson Lakes Campus, Adelaide, SA 5095, Australia
    Email: lakhmi.jain@unisa.edu.au

KES International
    2nd Floor, 145-157 St. John Street, London, EC1V 4PY, United Kingdom

Publisher
    Future Technology Press
    Virtual Knowledge Solutions Ltd
    PO Box 2115, Shoreham-by-sea,
    BN43 9AF, United Kingdom
    http://www.futuretechpress.com

ISBN 978-0-9561516-0-5

Foreword

Dedicated to the memory of Ignacio "Nacho" Barrio Moliner, researcher of the Soft Computing Group at Universitat Politècnica de Catalunya, who died prematurely from cancer in 2008.

Preface

Computational Intelligence (CI) is a useful, even if somehow ill-defined, umbrella term that covers a series of models and techniques mostly born out of the Artificial Intelligence domain, which loosely have in common their rejection of symbolic approaches. Connectionist models such as Artificial Neural Networks and Machine Learning techniques in general, Evolutionary Computation and Fuzzy Systems, amongst others, they all contribute to CI, which, in just a few years, has become the subject of a consolidated and rapidly-expanding IEEE Society.

AI methods are by now means newcomers to the field of medical applications in general and human oncology in particular. This type of research can be traced back to almost half a century ago. The 1970s and '80s witnessed a growing interest in the use of medical expert systems, yielding some success stories but also a number of disappointments, mostly put down to a mismatch between expert system-based medical DSS goals and the needs of health professionals working in real environments with tight requirements. Over the last two decades, CI methods have steadily colonized the research territory occupied by expert systems and permeated most areas of data-based medical research. Developments have focused on academic research, though, seldom reaching everyday medical practice. For this reason, it might be argued that the use of CI in clinical medicine has gone through an adolescent period of fast growth propelled by the excitement about its potential and novelty. This excitement has been tempered in more recent years by the need for generality, robustness and applicability of early exploratory findings. These needs provide researchers in CI with new and interesting challenges to face, as part of which they must reach to clinicians and the health systems they belong to in order to initiate fruitful collaborations, where key clinical questions are meant to be the main drivers of data-based studies.

Human cancer is one of the frontiers of medical research, and one in which CI has been very active over the last decade. As exemplified by the studies included in this book, CI research in cancer now reaches all biological scales from human population down to the phenotype and genotype. It also reaches the complete range of cancer pathologies: breast, gastrointestinal, genitourinary, gynecologic, head and neck, hematologic, lung, melanoma, prostate, and brain tumours, to name just the main categories. CI research in cancer faces some very specific challenges. To name a few: the own sensitive nature of the pathology when it comes to the provision of diagnosis and prognosis; the usually heterogeneous and incomplete information available; and the difficulty of creating homogeneous multi-centre cancer databases due to lack of unified standards and strict information privacy requirements.

This book does not aim at completeness in the sense of covering all cancer pathologies (the first three chapters of the book deal with breast cancer, the next three with brain tumours and the last two, in turn, with gynecologic and gastrointestinal cancer) and at all scales of medical research. Instead, the studies reported in the following chapters should provide readers with a broad palette and hopefully useful taster of what CI methods can achieve in the field.

Biographical Notes

Alfredo Vellido received his degree in Physics from the University of the Basque Country (Spain) in 1996. He completed his PhD at Liverpool John Moores University (U.K.) in 2000, and briefly joined Liverpool John Moores University again as a postdoc for a project in the field of computational neurosciences. He is now a Spanish MICINN *Ramón y Cajal* research fellow with the Technical University of Catalonia, in Barcelona. Research interests include pattern recognition, statistical machine learning and data mining, as well as their application in medicine, market analysis, ecology, and e-learning, on which subjects he has published widely. He currently leads the Spanish CICyT research project AIDTumour.

Paulo J.G. Lisboa graduated from the University of Liverpool with a Ph.D. in theoretical physics. He joined Liverpool John Moores University as Chair of Industrial Mathematics in 1996. He co-leads the Cancer SIG of the European Network of Excellence Biopattern and has served as expert evaluator and project reviewer for the European Commission. He has chaired the Healthcare Technologies Professional Network of the IEE and the AIME, and represents the IEE in the Royal Academy of Engineering's UK Focus for Biomedical Engineering. His interest in neural networks applied to control, image processing and medical spectroscopy led to several edited books and over sixty refereed publications.

Contents

Chapter 1
Investigating human cancer with Computational Intelligence techniques

Alfredo Vellido * and Paulo J. G. Lisboa

Abstract Driven by the growing demand of personalization of medical procedures, data-based, computer-aided cancer research in human patients is advancing at an accelerating pace, providing a broadening landscape of opportunity for Computational Intelligence methods and related techniques. This landscape can be observed from the wide-reaching view of population studies down to the genotype detail. In this introductory chapter, we provide a sweeping glimpse, by no means exhaustive, of the state-of-the-art in this field at the different scales of data measurement and analysis. We do so by focusing mostly on examples from European research, some of which are the matter of the following chapters of the book.

1.1 Introduction

The growing demands from ever better informed patients, with their increasingly sophisticated expectations from individual doctors, health institutions, and health systems as a whole, conform one of the forces driving the current upheaval towards personalized medicine. Opening the gates of medical information accessibility to the public in general will have effects that should not be underestimated. Perhaps unsurprisingly, even the use of the most pervasive of modern search engines, Google, for medical diagnosis assistance has been favourably evaluated [1] as an aid for expert diagnosis. What is striking is that the same study reports that "it seems that patients

Alfredo Vellido
Dept. de Llenguatges i Sistemes Informàtics - Universitat Politècnica de Catalunya, C. Jordi Girona, 1-3. 08034, Barcelona, Spain, http://www.lsi.upc.edu/~avellido, e-mail: avellido@lsi.upc.edu

* A. Vellido is a Spanish Ministry of Science and Innovation (MICINN) Ramón y Cajal program fellow researcher and acknowledges funding from CICyT project TIN2006-08114. Paulo J.G. Lisboa acknowledges funding from the Biopattern Network of Excellence FP6/2002/IST/1; N. IST-2002-508803.

use Google to diagnose their own medical disorders too". This raises the stakes for medical practitioners, who must face the challenge posed by savvy patients.

The true personalization of medical procedures is a difficult, if not unsurmountable, task, and balancing the improvement of health care delivery it entails and the corresponding escalating costs in all its phases -prevention, diagnosis, prognosis, and therapy- maybe beyond reach for almost any public and universal health system, especially in the area of oncology. Such balance maybe stricken, at least partially, with the introduction and application of new translational tools for medical knowledge discovery and computer-based decision support systems (DSS).

These tools rely heavily on evidence-based medicine (EBM) and have the potential to help in reducing unreliability and errors in all phases of care [2]. In fact, EBM can be thought of as a framework to deal with medical uncertainty, as its operational principles can be used to answer the questions for which there is good evidence and narrow the questions for which evidence is lacking [8]. This is the context in which Computational Intelligence (CI) and related methods can play a useful role in cancer research. Some issues concerning personalized medicine and CI are further discussed in section 2.

EBM has been defined as "healthcare practice that is based on integrating knowledge gained from the best available research evidence, clinical expertise, and patients' values and circumstances" [1], and its practice requires health management to be based on objective findings, rather than on beliefs or subjective interpretation of the knowledge-base. The evidence available to the medical practitioner can indeed take different and very heterogeneous forms, but we are specifically interested here in quantitative information, be it in the form of medical data records, patient's biological signal, etc. In any case, the evidence, both qualitative and quantitative, available to medical decision makers is growing exponentially, even in an area as specific as oncology. This is a mixed blessing: more evidence can lead to better informed decision-making in the form of diagnosis and prognosis of pathologies, but the heterogeneity of many available cancer databases and the amount of cancer-related published research can make it difficult for medical decision makers to extract useful and usable knowledge from it. This situation further justifies the design and development of computer-based DSS for the use of medical experts and lays the ground for the introduction of CI and related methods at their core. This is discussed, in more detail, in section 4.

In this enterprise, CI experts have to fight their ground, so to speak, first with its traditional tenants, the specialist medical statisticians. But even more importantly, they have to fight the ground of acceptance by medical institutions and practitioners. And this fight can be fierce: only if the opinion that computer- data-based medical DSS are both reliable and useful permeates the opinion of oncology experts and the public in general (and especially of cancer patients and organizations), are clinicians likely to embrace this tools in day-to-day practice. Therefore, the role of medical DSS, and especially of those involving somehow exotic methods such as the ones stemming from the CI field, should be carefully defined and discussed as part of their design. One way to ensure that the implementation of these technologies reaches enough critical mass is consolidating their development through international large-scale research projects. Some examples of these in the area of cancer research in Europe are described in section 3.

1.2 Personalized medicine and Computational Intelligence

It is not only better informed patients that are driving the progress towards personalized healthcare. Advances in medical science, biomedical engineering and molecular biology, reinforced by social attitudes centered on consumer choice also lead towards a system in which medical care is tailored to the specific needs of patients and their families. The agenda of personalized medicine is further driven by:

- Technological drives: The accelerating pace of science and technology which is opening-up new and compelling possibilities for healthcare development with a concomitant growth in public expectations.
- Economical drives: The cost of healthcare provision, in particular for chronic diseases and cancer, has generated a growing demand for ambulatory care, autonomous monitoring and control, and intelligent decision support for clinicians and patients alike.
- Legal drives: Litigation, in the increase especially in western health systems, which diminishes margins for human error, thus spurring greater reliance on technological assistance.

The issues highlighted above illustrate the potential engineering challenges that are expected to shape personalized medicine. Some of these challenges could be addressed by translational research to harness fundamental models from systems biology and pharmacogenomics that can result in clinical tools that are affordable and robust. The interface between molecular biology, wet biology, clinical progression, informatics, mathematical sciences and CI is now a focal point of highly interdisciplinary research with high expectations for delivery of radical insights into the progression of cancer pathologies.

An example of this is the recent discovery of the possibility that particular biological pathways cause tumors to metastasise very early in their development, simultaneously signalling to prevent metastatic growth. If even small tumors with this pathology are excised, this will activate the dormant micro-metastases leading to a generalized cancer within months of surgery. Detailed non-linear statistical analyses (using a precedent of wavelet analysis for hazard rate estimation) of prospectively acquired longitudinal data over long periods of follow-up identified the presence of early, unexpected peaks, from which this hypothesis was first generated and then tested in animal models [5]. Current research aims to characterize these patients from large integrated data sets of clinical and bioinformatics data so as to identify the characteristic profile for a group of patients who may, in principle, be saved by carefully timed adjuvant therapy. This example shows how the interface between advanced methods in different disciplines can challenge widely accepted models of care, in this case the possibility that tumor surgery may, of itself, have deleterious effects.

1.3 Computational Intelligence for cancer research in current European research projects

There are many ways in which we could have structured the information related to the application of CI methods in human cancer research in this introductory chapter [6, 9]. We could, for instance, have organized it according to the broad palette of CI methods available, or perhaps according to the categorization of tumor pathologies as provided by the World Health Organization (WHO) system. Here, we take a different and less obvious point of view and follow a top-down hierarchy of clinical measurement scales. This information is summarized in Table 1.1, emphasizing, at each scale, the relevance to CI methods.

All the chapters in this book report research carried out in the European context. The European Commission Information and Communication Technologies (ICT) for Health Unit of the Information Society and Media Directorate General, manages a series of international research projects in the medical ambit, funded under the Sixth Framework program (FP6). The program's 4th call, "Integrated biomedical information for better health", involves the "research and development on innovative ICT systems and services that process, integrate and use all relevant biomedical information aimed for improving health knowledge and processes related to prevention, diagnosis, treatment, and personalization of health care". Here, biomedical information would include "not only clinical information relating to tissues, organs or personal health-related information, but also information at the level of molecules and cells, such as that acquired from genomics and proteomics research" [8].

Within this 4th call, several projects concern cancer research more or less directly, namely: Advancing Clinico-Genomic Clinical Trials on Cancer (ACGT), Association Studies Assisted by Inference and Semantic Technologies (ASSIST), Computational Intelligence for Biopattern Analysis in Support of eHealthcare (Biopattern), Agent-based Distributed Decision Support System for Brain Tumour Diagnosis and Prognosis (HealthAgents), Structured European Biomedical Informatics to Support Individualised Healthcare (INFOBIOMED), and Automated Diagnosis System for the treatment of Colon Cancer by discovering mutations on tumour suppressor genes (MATCH). All these projects involve, in one way or another, data analysis, and some of them realize it through data mining or CI methods, often related to machine learning. In Table 1.2, we can find a summary of the general goals of some of these projects, including more specific data analysis goals and their use of CI and related methods.

Scale	Clinical drivers	Study types	Relevance to CI methods
Population	Integrated support for patients.	Epidemiology	Formal acquisition and validation platforms: data standards and clinical protocols for prospective and retrospective studies.
Individuals (system level)	Disease Biology and Computer-Aided Decision Support.	Personalised Decision Support with Interfaces for Clinicians and Patients; Personalised Monitoring; and Point of Care Diagnostics.	Predictive diagnostics and rule generation for translational research
Clinical signs (organ level)	Disease Biology, Computer-Aided Decision Support, Response to Therapy, and Tumour Delineation.	Personalised Decision Support with Interfaces for Clinicians and Patients; Personalised Monitoring; and Point of Care Diagnostics	Longitudinal data analysis for prognostic modelling
Physiological measurement (tissue level, $10^{-3} - 10^{-2}$m)	Disease Biology, Computer-Aided Decision Support, Response to Therapy, and Tumour Delineation.	Multimodal Data Fusion; Virtual Physiological Human.	Fusion of anatomical and functional data including different measurement modalities
Immunohistochemistry (cell level, $10^{-5} - 10^{-4}$m)	Disease Biology, Computer-Aided Decision Support, Response to Therapy, and Tumour Delineation.	Integration with higher level data linked to clinical expertise about the pathology, and Pharmacogenomics.	High-dimensional data modelling for diagnostic and laboratory imaging
Phenotype (pathways, $10^{-7} - 10^{-6}$m)	Disease Biology, Computer-Aided Decision Support, Response to Therapy, and Tumour Delineation.	Integration with higher level data linked to clinical expertise about the pathology, and Pharmacogenomics.	Knowledge discovery from data, clustering and visualization
Genotype (genes, $10^{-9} - 10^{-8}$m)	Disease Biology, Computer-Aided Decision Support, and Response to Therapy.	Exploratory data analysis for hypothesis generation.	Modelling protein networks, with fewer observations than potential predictors

Table 1.1 Overview of the role of data analysis and CI methods in human cancer research.

Project name	General goals	Use of CI and related methods
ACGT	It aims to fill-in the technological gaps of clinical trials for two pathologies: breast cancer and paediatric nephroblastoma. The project will develop a Biomedical GRID infrastructure supporting mediation services for sharing clinical and genomic expertise, helping to identify what determines which form of treatment suits which patient.	Data mining tools, using R language in a grid environment, including exploratory analysis using SOM, k-Means and Sammon's mapping in combination with statistical techniques. These are combined with tools for interactive visualization. Some sort of data mining methodological framework, understood as "workflow mining" is also provided.
ASSIST	The project aims to provide medical researchers of cervical cancer with an environment that will unify multiple patient record repositories. They will be able to combine phenotypic and genotypic data and perform association studies on larger multi-center sets of patient records.	It claims that, given a hypothesis that needs validation, the system must be able to process relevant records and mine the collected data. Mixed text and data mining are integrated in a workflow process using weighted fuzzy methods, neural networks, and support vector machines.
Biopattern	This project's goal is to develop a pan-European, intelligent analysis of a citizen's bioprofile; to make the analysis of this bioprofile remotely accessible to patients and clinicians; and to exploit bioprofile to combat major diseases such as cancer and brain diseases. It includes studies on ovarian, breast and brain cancers, leukaemia and melanoma.	It proposes to provide online novel computational intelligent techniques for a pan-European integrated analysis of an individual's bioprofile. They include artificial neural networks (ANNs), evolutionary algorithms, support vector machines, Bayesian methods, Adaptive Resonance Theory, Tree models, Fuzzy techniques, etc.
Health Agents	This project plans to create a multi-agent distributed DSS for the early diagnosis and prognosis of brain tumours. A distributed Data Warehouse with the world's largest network of interconnected databases of clinical, histological, and molecular phenotype data of brain tumour patients will be created.	It aims to develop new pattern recognition methods for a distributed classification and analysis of HR-MAS (High-Resolution under Magic-Angle Spinning) and microarray data. They resort to feature selection, clustering and classification techniques, including support vector machines, mixture models and Bayesian methods.

Table 1.2 Several current European research projects on cancer and their use of CI techniques.

1.4 What can Computational Intelligence methods do for cancer research?

1.4.1 Managing the complexity of medical data

Cancer data are often not acquired with the specific purpose of data-based modelling. This means that the data available to the analyst can be inconsistently recorded and may also include substantial missing information. Inconsistency might arise, for instance, from the difficulty of merging heterogeneous information: quantitative data of different types and perhaps measured over time or at different times, including complex multivariate biosignals; qualitative data in textual form, etc. An area in which this can be a major limitation is cancer genomics [9]. Inconsistency might also have a cause that is more difficult to address: the existence of different clinical protocols for different clinical centers and, perhaps, even different regional or national guidelines, not to mention non-standard technical specifications of measurement devices. Even when the analysed data are the result of careful pre-processing and thoroughly checked for consistency, CI methods can still reveal data limitations. In this book, MRS data corresponding to human brain tumors were explored using outlier detection, visualization and rule extraction techniques in [10, 11]. These data had previously undergone a painstakingly detailed depuration process [12] and, even though, this study pinpointed atypical data cases whose presence would worsen the analysis using CI methods.

It has been acknowledged [13] that a lack of a common infrastructure has largely prevented the analysis of disparate, multi-level data sources, the result being the realization of few cross-site studies and multi-centric clinical trials, making the integration of multi-level data difficult (which does not preclude single-center studies, such as [14] in this book, providing valid and insightful new knowledge). Decision support will have to resolve the mathematical challenges around fusion of multi-modal data which together make-up the individual's bioprofile, where the health-related data it comprises is distributed over time, points of care and pathological domains, comprising genetic markers through physiological measurements and clinical signs. Some of the European research projects described in the previous section, namely ACGT, Biopattern and HealthAgents deal actively with some of these issues.

Despite being often overlooked, the issue of missing data may be dealt with using statistical and CI techniques. Within the research of CI methods in oncology, the impact of missing data in evaluating ANNs trained on complete breast cancer data was assessed in [15]. Missing data imputation using a Statistical Machine Learning (SML) model was considered in [4] for a problem of brain tumour data exploration.

An often quoted advantage of many CI methods is that they have the potential to scale well to large databases. It has been stated [13] that in areas such as cancer diagnostics "the data gathering capabilities have greatly surpassed the data analysis techniques". For these authors, the "ultimate challenge in coming years" will be the automation of the processes of knowledge discovery from this data, which is the realm of CI methods. In fact, one of the main niches for CI is found in cancer pathologies requiring the combination of multiple, or complex forms of data [17, 18].

1.4.2 Robustness of modelling methodology

There have been numerous calls to reinforce the robustness of CI models by applying to them established statistical frameworks for complexity control and so to combat common problems such as data overfitting [19]. Sound statistical principles are essential to trust the evidence base built with any computational analysis of medical data [20]. This is clearly pointed out in [21], when stating that "in the context of gene expression analysis a main goal, besides classification, is finding a gene signature [...]. When dealing with high throuput data the choice of a consistent selection algorithm is not sufficient to guarantee good results. It is therefore essential to introduce a robust methodology to select the significant variables not susceptible of selection bias and to use valid statistical indicators to quantify and assess the significance of the results".

This is true for AI-based models in general and CI methods in particular. This would naturally lead to the recommendation to use SML methods in the area, which are already establishing themselves in the more general field of bioinformatics [22]. This is not say that sophisticated SML methods should be considered the adequate modelling choice by default, as counterexemplified by the experiments reported in one of the following chapters [23]. An example of SML method that can be found in this book is the Least-Squares SVM used in [24] for the classification of different types of ovarian tumors. Also in this book, and addressing method robustness from a different point of view, Etchells and colleagues [25] propose a new methodology to improve the robustness of a statistical and neural network prognostic study of longitudinal data from patients with operable breast cancer.

1.4.3 Demonstrating clinical benefit

It is also been reported that there are relatively few published proper tests showing the clinical value of ANNs against established linear-in-the-parameters statistical methods [20]. In this book, this type of comparison is nicely illustrated by the experiments in [25]. This potential shortcoming could be extended to other CI methods and become a crucial bottleneck for their adoption in real medical decision making, given that, in critical fields such as oncology data-based decision making, "techniques lacking a solid foundation are not readily accepted" [26]. This "lack of solid foundation" may be in the eye of the beholder, and much might be argued for the maturity of many CI techniques. At this point, the methods for the evaluation of CI techniques must also be carefully chosen. The use of Receiver Operating Characteristic (ROC) analysis for model evaluation has been recommended by, amongst others, [20, 15].

1.4.4 Model transparency

One of the potential drawbacks affecting the application of CI methods in general is the usually limited interpretability of the results they yield. This is again an extremely sensitive issue in a critical context such as clinical oncology. One way to overcome this limitation is by explaining the operation of CI models using rule extraction methods. The factor of interpretability should not be underestimated in the real medical practice where "most physicians are not even accustomed to the idea of computer-aided problem solving" [26]. Several authors have, in recent years, resorted to rule extraction from CI models in cancer research. Many of these involve the analysis of breast cancer data [28, 29, 9], perhaps because of the ready availability of sufficiently large standardized databases, although rule extraction from the classification of other cancer pathologies has also been implemented. An example of this in this book can be found in the work of Nebot et al. [11], where the simple and interpretable obtained classification rules are able to faithfully describe different brain tumor pathologies in terms of a very reduced subset of spectral data frequencies.

General best practice recommendations for the use of ANNs in cancer research, including training-validation-testing techniques, evaluation of performance and reliability, calculation of confidence intervals and misclassification costs, feature selection and extraction, and model regularization, can be found in [20, 18]. For further cautionary tales on the use (and misuse) of ML methods in oncology, we refer the reader to [31, 32].

Many biomedical domains related to cancer, such as genomics, metabolomics, medical imaging and spectrometry make use of very high-dimensional data, and their analysis and interpretation usually require dimensionality reduction techniques. Again, we can find some examples of the importance of feature selection for dimensionality reduction, and how it improves model transparency, in this book. A study that set out to classify MRS data corresponding to human brain tumors [33] found that a parsimonious selection of data frequencies was better at the given classification task than the full set of hundreds of spectral frequencies. In [27], a new method of probe selection is proposed in the context of DNA microarray analysis for a problem of prediction of breast tumor sensitivity to pre-operative chemotherapy.

An alternative approach to achieve model transparency is the use of graphical models for the exploration of relationships between data features and given outcomes. As an example, in [34] Bayesian Networks were used for the exploration of such relationships in Nasopharyngeal Carcinoma occurrence in Maghrebian countries.

1.5 Conclusion

Recent reports of EUROCARE, a research collaboration established in 1989 and now in its 4th stage, currently involving population-based cancer registries from 23 European countries, highlight that substantial changes in cancer patient survival have arisen in recent years, reflecting changes in the health care systems, implementation of national cancer plans, implementation of screening programmes for new cancers and, importantly, progress in diagnosis and therapeutics. The latest data [35] indicate that the gap between European countries in cancer survival is narrowing, suggesting substantial improvement in cancer care in countries with poor survival.

One of the keys to the improvement of cancer survival figures is discovery-driven translational research and, as part of it, the development of efficient medical DSS for the support of medical diagnosis and prognosis of pathologies. CI can provide, coalescing with more traditional statistical approaches, robust methods to be deployed at the core of these DSS, ready to properly manage uncertainty within the EBM framework. In order to pass from adolescence to fruitful maturity, CI studies of medical data must broaden their focus beyond technical detail to pay greater attention to medical requirements. Over the last decade, clinicians and medical researchers have become more aware of what these methods are and what can they achieve, so that closer collaboration with them should help the data analyst to drive data-based studies according to key clinical questions, therefore building into study design relevance. A further challenge that CI practitioners will have to face in coming years is that of data management, as multi-centre and international databases become more of a standard, while the available biomedical signal for analysis becomes increasingly multivariate, multiscale and multimodal [36].

References

1. Tang, H., Ng, J.H.K.: Googling for a Diagnosis-Use of Google as a Diagnostic Aid: Internet Based Study. Brit. Med. J. 333, 1143–1145 (2006)
2. Conforti, D., Tsiknakis, M., Habetha, J., Bacquet, J.: Introduction to the HEALTHINF Special Session on Knowledge Discovery and Decision Support Systems in Health Information Systems. *International Joint Conference on Biomedical Engineering Systems and Technologies* (BIOSTEC 2008), INSTICC Press, January 28-31, Funchal (Portugal), 2008.
3. Ghosh, A.K.: On the Challenges of Using Evidence-Based Information: The Role of Clinical Uncertainty, J. Lab. Clin. Med. 144(2), 60–64 (2004)
4. Dickersin, K., Straus, S.E., Bero, L.A.: Evidence Based Medicine: Increasing, not Dictating, Choice. Brit. Med. J. 334(supl.1) s10 (2007)
5. Demicheli, R., Bonadonna, G., Hrusheksy, W.J.M., Retsky, M.W., Valagussa, P.: Menopausal Status Dependence of the Timing of Breast Cancer Recurrence after Surgical Removal of the Primary Tumour, Breast Cancer Res. 6 R689-R696 (2004)
6. Vellido, A., Lisboa, P.J.G.: Neural Networks and Other Machine Learning Methods in Cancer Research. In Procs. of IWANN 2007, LNCS Vol. 4507, 964–971 (2007)
7. Vellido, A., Biganzoli, E., Lisboa, P.J.G.: Machine Learning in Cancer Research: Implications for Personalised Medicine. In: 16th European Symposium on Artificial Neural Networks (ESANN 2008), d-Side pub., Evere, Belgium. 55–64 (2008)

8. ICT for Health Unit: Resource book of eHealth projects. Working Document, European Commission Information Society and Media DG, European Communities, 2006.
9. Louie, B., Mork, P., Martín-Sánchez, F., Halevy, A., Tarczy-Hornoch, P.: Data Integration and Genomic Medicine, J. Biomed. Inform. 40, 5–16 (2007)
10. Vellido, A., Julià-Sapé, M., Romero, E., Arús, C.: Exploratory Characterization of a Multi-Centre ^{1}H-MRS Brain Tumour Database. In current volume (2008)
11. Nebot, À., Castro, F., Vellido, A., Julià-Sapé, M., Arús, C.: Rule-Based Assistance to Brain Tumour Diagnosis Using LR-FIR. In current volume (2008)
12. Julià-Sapé, M., Acosta, D., Mier, M., Arús, C., Watson, D., The INTERPRET Consortium: A Multi-Centre, Web-Accessible and Quality Control-Checked Database of in Vivo MR Spectra of Brain Tumour Patients. Magn. Reson. Mater. Phy. MAGMA 19, 22–33 (2006)
13. Tsiknakis, M., Sfakianakis, S., Rueping, S., Trelles, O., Siestang, T., Claerhout, B., Virvilis, V.: A Semantic Grid Services Architecture in Support of Efficient Knowledge Discovery from Multilevel Clinical and Genomic Datasets. In L. Azevedo and A.R. Londral, editors, Proceedings of the 1st First International Conference on Health Informatics (HEALTHINF 2008), Vol.1, INSTICC Press. 279–287 (2008)
14. Distaso, A., Abatangelo, L., Maglietta, R., Creanza, T.M., Piepoli, A., Carella, M., D'Addabbo, Luca A., Mukherjee, S., Ancona, N.: Statistical Assessment of MSigDB Gene Sets in Colon Cancer. In current volume (2008)
15. Markey, M.K., Tourassi, G.D., Margolis, M., DeLong, D.M.: Impact of Missing Data in Evaluating Artificial Neural Networks Trained on Complete Data, Comput. Biol. Med. 36, 516–525 (2006)
16. Vellido, A., Lisboa, P.J.G.: Handling Outliers in Brain Tumour MRS Data Analysis through Robust Topographic Mapping. Comput. Biol. Med. 36, 1049–1063 (2006)
17. Lisboa, P.J.G., Taktak, A.F.G.: The Use of Artificial Neural Networks in Decision Support in Cancer: a Systematic Review, Neural Networks 19, 408–415 (2006)
18. Dayhoff, J.E., DeLeo, J.M.: Artificial Neural Networks: Opening the Black Box, Cancer 91, 1615–1635 (2001)
19. Ripley, B.D., Ripley, R.M.: Neural Networks as Statistical Methods in Survival Analysis. In R. Dybowski and V.Gant, editors, Clinical Applications of Artificial Neural Networks, Cambridge University Press, Cambridge. 237–255 (2001)
20. Lisboa, P.J.G.: A Review of Evidence of Health Benefit from Artificial Neural Networks in Medical Intervention, Neural Networks 15, 9–37 (2002)
21. Barla A., Mosci, S., Rosasco, L., Verri, A.: A Method for Robust Variable Selection with Significance Assessment. In M. Verleysen, editor, Proceedings of the 16th European Symposium on Artificial Neural Networks (ESANN 2008), d-side pub., Evere, Belgium. 83–88 (2008)
22. Baldi, P., Brunak, S.: Bioinformatics: The Machine Learning Approach, MIT Press, Cambridge, 2001.
23. Bacciu, D., Biganzoli, E., Lisboa, P.J.G., Starita, A.: Unsupervised Breast Cancer Class Discovery: a Comparative Study on Model-based and Neural Clustering. In current volume (2008)
24. Daemen, A., Gevaert, O., Leunen, K., Vanspauwen, V., Michils, G., Legius, E., Vergote, I., De Moor, B.: A Genome-Wide Computational Study of Copy Number Variations: an Example on Ovarian Cancer. In current volume (2008)
25. Etchells, T.A., Fernandes, A.S., Jarman, I.H., Fonseca, J.M., Lisboa, P.J.G.: Stratification of Severity of Illness Indices and Out-of-Sample Validation: A Case Study for Breast Cancer Prognosis. In current volume (2008)
26. Lucas, P.J.F.: Model-Based Diagnosis in Medicine, Artif. Intell. Med. 10, 201–208 (1997)
27. Natowicz, R., Incitti, R., Rouzier, R., Çela, A., Braga, A., Horta, E., Rodrigues, T., Costa, M., Pataro, C.D.M.: A New Method of DNA Probes Selection and its Use with Multi-Objective Neural Network for Predicting the Outcome of Breast Cancer Preoperative Chemotherapy. In current volume (2008)
28. Tan, K.C., Yu, Q., Heng, C.M., Lee, T.H.: Evolutionary Computing for Knowledge Discovery in Medical Diagnosis, Artif. Intell. Med. 27, 129–154 (2003)

29. Etchells, T.A. et al.: Empirically Derived Rules for Adjuvant Chemotherapy in Breast Cancer Treatment. In IEE Proceedings of the 2nd International Conference on Advances in Medical Signal and Information Processing (MEDSIP). 345–351 (2004)
30. Hassanien, A.E.: Fuzzy Rough Sets Hybrid Scheme for Breast Cancer Detection, Image Vision Comput. 25, 172–183 (2007)
31. Schwartzer, G. et al.: On the Misuses of Artificial Neural Networks for Prognostic and Diagnostic Classification in Oncology, Stat. Med. 19, 541–551 (2000)
32. Cruz, J.A., Wishart, D.S.: Applications of Machine Learning in Cancer Prediction and Prognosis, Cancer Informatics 2, 59–78 (2006)
33. Gonzlez, F.F., Belanche, L.A.: Feature and Model Selection in ^{1}H-MRS Single Voxel Spectra for Cancer Classification. In current volume (2008)
34. Rodrigues de Morais, S., Aussem, A., Corbex, M.: Handling Almost-Deterministic Relationships in Constraint-Based Bayesian Network Discovery: Application to Cancer Risk Factor Identification. In M. Verleysen, editor, Proceedings of the 16th European Symposium on Artificial Neural Networks (ESANN 2008), d-side pub., Evere, Belgium. 101–106 (2008)
35. Berrino, F., De Angelis, R., Sant, M., Rosso, S., Lasota, M.B., Coebergh, J.W., Santaquilani, M., and the EUROCARE Working group: Survival for Eight Major Cancers and All Cancers Combined for European Adults Diagnosed in 1995-99: Results of the EUROCARE-4 Study, Lancet Oncol. 8, 773–783 (2007)
36. Cerutti, S.: Multivariate, Multiorgan and Multiscale Integration of Information in Biomedical Signal Processing. Keynote lecture. In Proceedings of the 1st International Joint Conference on Biomedical Engineering Systems and Technologies (BIOSTEC 2008), INSTICC Press, pages IS5–IS8 (2008)

Chapter 2
Unsupervised Breast Cancer Class Discovery: a Comparative Study on Model-based and Neural Clustering

Davide Bacciu, Elia Biganzoli, Paulo J. G. Lisboa, and Antonina Starita

Abstract CoRe learning, a novel neural network clustering algorithm, is applied to the unsupervised tumor class discovery on a breast cancer dataset. The tumoral subgroups discovered by CoRe learning are compared with previously published results obtained by Partition Around Medoids, as well as with the sample partitions obtained by model-based algorithms grounding on Gaussian mixtures and Bayesian methods. It is found that CoRe has the highest agreement with the previously published results, while model-based algorithms cannot determine a shared consensus on a stable partition of the samples. Moreover, CoRe supports a prognostic hypothesis that postulates the existence of an additional tumor subgroup with respect to previously published results. Such an additional cluster is hypothesized to differentiate samples with intermediate progesterone and estrogen expression into treatment resistant and hormone-responsive cases.

2.1 Introduction

The unsupervised discovery of the processes underlying biological phenomena is an open challenge for the computational intelligence community that has been addressed by several unsupervised learning models with roots in statistics, fuzzy logic and neural networks, just to mention a few. Cancer research, and in particular breast tumor research, is one of the areas that has profited more of computational intelligence techniques for biomedical data exploration [1]. In particular, unsupervised learning models have been used to estimate the latent structure of the data, discovering homogenous groups of samples that are hypothesized to correspond to distinct tumor subtypes.

Davide Bacciu
IMT Lucca Institute for Advanced Studies, P.zza San Ponziano 6, 55100 Lucca, Italy, e-mail:
d.bacciu@imtlucca.it
Dipartimento di Informatica, Università di Pisa, Largo B. Pontecorvo 3, 56127 Pisa, Italy

In this work, we address the unsupervised discovery of prognostic classes in breast cancer data along with issue of estimating the relevance of the features describing the data samples. We compare a neural clustering algorithm, that is Competitive Repetitions Suppression learning (CoRe), with sophisticated Bayesian approaches relating to Gaussian Mixture Models. CoRe learning is a soft-competitive model inspired by a memory mechanism of the visual cortex, named Repetition Suppression (RS). The RS mechanism induces long-lasting changes to the visual cortex, decreasing the neural activity as a consequence of the repeated presentation of similar stimuli. In brief, it produces a sharpening of the neural representation of items by means of an overall reduction of the number of active neurons, which is counterbalanced by the steepening of the response of the most item-selective neurons. This process seems to be aimed at the selection of neurons that act as detectors of the most informative features. The CoRe model has first been applied in [2] to the cluster number identification problem and it has been extended recently in [3] to deal with high-dimensional data and with feature ranking. We apply this latter CoRe extension to analyze a case-study dataset from breast cancer research [4] with the intent of discovering clusters of functionally correlated samples, extracting cancer profiles characterized by different tumoral dynamics. Moreover, CoRe's feature ranking is used to gather insight into the markers that best describe the discovered bio-profiles.

Before reporting the experimental results, we introduce an alternative CoRe formulation that highlights its interpretation as a competitive neural network model, analyzing similarities and differences with unsupervised neural learning algorithms such as, for instance, Self Organizing Maps (SOM) [5], Growing Neural Gas (GNG) [6] and Adaptive Resonance Theory (ART) [7].

2.2 Neural Clustering with Competitive Repetition-Suppression Learning

A CoRe network is a two-layer neural network (see Fig. 2.1) where input nodes are fully connected to a layer of output units $L^o = \{u_1^o, \ldots, u_f^o\}$ that compete with each other through lateral connections. Each output unit u_i^o is associated to a prototype vector $c_i \in \mathbb{R}^d$, determining its preferred stimulus, and to an activation function $\varphi_i(x_k)$ that determines the unit's response to the input pattern $x_k \in \mathbb{R}^d$. In other words, c_i and φ_i together determine the location and shape of the neuron's receptive field. In the remainder of the paper we will focus on univariate Gaussian activation functions φ_i centered on c_i and parametrical with respect to the spread σ_i. For each of the l-th components of the d dimensional input vector, this defines an activation $\varphi_{il}(x_{kl}) \in [0, 1]$. This property will be used in the following to selectively suppress the irrelevant components of the prototype vector c_i.

The units in the outer layer are fully connected through a set a lateral inhibitory connections that serve to convey the suppressive potential to the loser neurons. More in detail, for an input pattern x_k we first calculate the activation $\varphi_i(x_k)$ of each output

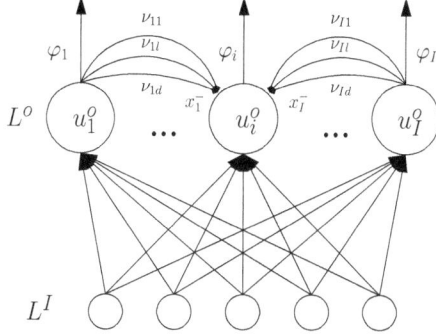

Fig. 2.1 CoRe neural network with lateral inhibition: intra-layer connections link the competitive neurons u_i^o and propagate the Repetition Suppression. Lateral inhibition is applied selectively to the single input components l, i.e. there are d lateral connections from neuron u_j^o to neuron u_i^o, each weighted with a (possibly) different v_{il}^t.

neuron u_i^o. Then, the most active units are selected to form the *winners pool*, i.e.

$$win_k = \{i \mid \varphi_i(x_k) \geq \theta_{win}, \ u_i^o \in L^o\} \cup \{i \mid i = \arg\max_{j \in L^o} \varphi_j(x_k)\}, \qquad (2.1)$$

while the remainder of the neurons is inserted into the *losers pool*, that is

$$lose_k = \{i \mid \varphi_i(x_k) < \theta_{win}, \ u_i^o \in L^o\} \setminus \{i \mid i = \arg\max_{j \in L^o} \varphi_j(x_k)\}. \qquad (2.2)$$

Here θ_{win} determines the minimum activation level for winner units and, consequently, regulates the target selectivity of the neurons. The threshold θ_{win} does not directly determine the cluster number estimate, whereas it regulates the tradeoff between soft and hard competition among the neurons. Experimentally we have determined that the best results, for a generic clustering task, are obtained for $\theta_{win} \in [0.8, 1)$: lower θ_{win} values might produce slower convergence and incorrect cluster number estimation. Notice that the second term in (2.1) prevents the winner pool from being empty if no unit fires more than θ_{win}. Similarly, the second term in (2.2) discounts the most active unit from the loser pool in order to ensure that $\{win_k, lose_k\}$ is a partition of the units set L^o.

Each unit u_i^o generates an inhibitory output x_{il}^- for each of the l components of the d dimensional input, that is

$$x_{il}^- = \begin{cases} \varphi_{il}(x_{kl})v_{il}^t & \text{if } i \in win_k \\ 0 & \text{if } i \in lose_k \end{cases} \qquad (2.3)$$

where $0 \leq v_{il}^t \leq 1$ (i.e. the *stimulus predominance* in [3]) represents the weight of the inhibitory connection for the l-th component (see lateral links in Fig. 2.1). The inhibitory weight vector $v_i^t \in \mathbb{R}^d$ essentially measures the frequency of the patterns that are preferred by the i-th unit. The value of its l-th component (at time t) is computed as

$$v_{il}^t = \frac{1}{|\chi_t|} \sum_{x_k \in \chi_t} \min\left(\frac{\varphi_{il}(x_{kl})}{\varphi_{z(L^0,x_k)l}(x_{kl})}, 1\right) \qquad (2.4)$$

where χ_t is the set of patterns x_k presented to the network up to time t, $z(L^0,x_k)$ returns the index of most active unit for the pattern x_k and the function $\min(\cdot,1)$ is used to enforce the condition $v_{il}^t \leq 1$. In general, the denominator $\varphi_{z(L^0,x_k)l}(x_{kl})$ acts a normalization factor and can be replaced with other suitable terms such as, for instance, the total feature activation $\sum_j \varphi_{jl}(x_{kl})$. Notice that, by using this latter normalization, there is no more need of thresholding v_{il}^t by means of the $\min(\cdot,1)$ function, whose argument becomes the well known soft-max function.

The inhibitory output is propagated to all the competing neurons through the lateral connections and is accumulated at each unit i, yielding

$$RS_{il}^t(x_k) = \frac{1}{|win_k|} \sum_{j \in I_i^-} x_{jl}^- \qquad (2.5)$$

that is the l-th component of the *repetition suppression* generated at time t for the pattern x_k. The term $|win_k|$ represents the cardinality of the winners pool for x_k and is used to ensure $0 \leq RS_{il}^t \leq 1$, while I_i^- is the set of the inhibiting connections for the i-th neuron. In a sense, the RS_{il} term can be considered as the *net* value of the inhibiting inputs in the typical notation of the artificial neural network literature. The l-th component of the prototype vector c_i is updated as follows

$$\triangle c_{il}^t = \alpha_c \left[\delta_{ik} - (1 - \delta_{ik})\varphi_{il}(x_{kl})(RS_{il}^t(x_k))^2\right] \varphi_{il}(x_{kl})(x_{kl} - c_{il}^{t-1}) \qquad (2.6)$$

where α_c is the learning rate, while δ_{ik} is the indicator function for the winners pool, that is $\delta_{ik} = 1$ if $i \in win_k$ and $\delta_{ik} = 0$ otherwise.

Equations (2.5) and (2.6) essentially state that CoRe inhibition produces a penalization that is proportional to the frequency of the stimuli similar to the current input x_k. This penalization is applied only to the loser neurons, producing a displacement of the prototype vector components c_{il}^t with respect to the current input vector x_{kl} that is proportional to the amount of repetition suppression produced. Conversely, winner neurons are attracted by the current input and their activation is strengthened by moving the prototype components c_{il}^t towards x_{kl} proportionally to the respective activation levels $\varphi_{il}(x_{kl})$.

The formulation in (2.6) can be used to compare CoRe to other competitive learning models in literature. Consider, for instance, a typical prototype update function for a generic competitive neural network, that is

$$\triangle c_{il}^t = \alpha_c h(x_k, L)(x_{kl} - c_{il}^{t-1}) \qquad (2.7)$$

where $h(x_k, L)$ is a function regulating the type of competition between the units in the layer L. For instance, in hard competitive learning $h()$ is 1 for the best matching unit (BMU) and 0 otherwise, while in soft-competitive learning [8] $h()$ modulates the adaptation based on the firing level of the neuron. In SOM [5] and GNG [6], on the other hand, $h()$ is a *neighborhood* function determining the strength of prototype adaptation based on the proximity of the units to the BMU with respect to a given lattice, hence generating spatially ordered maps of the input stimuli. Comparing (2.7) with (2.6) suggests that CoRe's $h()$ function dynamically determines the strength and direction of adaptation based on whether the input pattern has already a sharp neural representation in the network.

It is interesting to note how each CoRe unit can be interpreted as a class detector, assigning positive class information to those patterns falling inside the excitatory region of its receptive field and negative class labels to those patterns in the inhibitory region. Moreover, it is worth noting that CoRe's lateral inhibition plays a substantially different role with respect to the typical lateral connections in competitive learning. The latter ones are often used as part of the competitive mechanism to determine the winner units for a given input pattern, but they do not convey a learning signal to the neurons. ART, for instance, uses the lateral inhibition to selectively shut-off committed neurons, allowing other nodes in the network to win the competition. CoRe, on the other hand, exploits the intra-layer links to propagate a teaching signal that produces a *long-term silencing* of the neurons and a suppression of the irrelevant input components. CoRe's lateral connections have somewhat much more in common with SOM lateral connections, that are indeed used to convey positive learning to other nodes of the network, although the two models differ significantly in the way they determine the neurons that receive teaching signals through the lateral connections (e.g. topographic proximity in SOM and activation sharpness in CoRe).

The prototype update rule in (2.6) can be obtained by differentiating the CoRe error

$$E_{il,k}^t = \delta_{ik}(1 - \varphi_{il}(x_{kl})) + (1 - \delta_{ik})\frac{1}{2}(\varphi_{il}(x_{kl})RS_{il}^t(x_k))^2 \qquad (2.8)$$

with respect to c_{il}^t (see [3] for further details). The same approach can be taken for all the variables in the activation function. For instance, differentiating the Gaussian φ_i with respect to the spread σ_{il} leads to the following update rule

$$\triangle\sigma_{il}^t = \alpha_\sigma \left[\delta_{ik} - (1 - \delta_{ik})\varphi_{il}(x_{kl})(RS_{il}^t(x_k))^2\right]\varphi_i(x_{kl})\frac{(x_{kl} - c_{il})^2}{\sigma_{il}^3}. \qquad (2.9)$$

The spread adjustment process in (2.9) adaptively reduces the variance of winner units, thus making them more selective. This behavior, on the one hand, complies with the biological repetition suppression phenomenon and, on the other hand, ensures the convergence of the learning process as described in [8] for soft competitive learning. CoRe's spread adjustment has also a self-regulating effect on frequent winner neurons. For instance, if a unit u_i^o wins for many samples, then its spread will reduce up to a point where u_i^o stops prevailing for certain patterns x_k; conversely, weakly winning units maintain wider spreads, retaining the possibility to compete on those x_k that are no more encoded by u_i^o.

CoRe learning tackles the fundamental challenge of generating compact neural coding of the input data: to evaluate the sharpness of the representation CoRe defines a metric for identifying the most significant units produced by the learning process as well as the most relevant components of the input vectors. The *relevance factor* for the l-th feature of the i-th neuron is modeled as

$$\hat{v}_{il}^t = \frac{1}{v_{il}^t |\chi_t|} \Sigma_{x_k \in win_{u_i^0}^t} \left\{ \frac{\varphi_i(x_{kl})}{\varphi_{z(win_k, x_k)l}(x_{kl})} \right\}_0 \quad \text{s.t.} \quad \{v\}_0 = \begin{cases} 0 & v > 1 \\ v & v \leq 1 \end{cases} \quad (2.10)$$

where $win_{u_i^0}^t$ is the set of patterns $x_k \in \chi_t$ for which unit u_i^0 was in the winners pool, while the function $\{\cdot\}_0$ flattens its argument to zero if it exceeds 1. The rationale behind this choice is to penalize the relevance of the l-th component of a prototype c_i whenever it produces an high feature activation $\varphi_{il}(x_{kl})$ in correspondence to a low feature activation $\varphi_{jl}(x_{kl})$ in the unit u_j^o that is the maximally active neuron for the pattern x_k, i.e. $j = z(win_k, x_k)$. This process implicitly considers improper responses all those feature activations that do not correspond to highly active features in the best matching unit. In this sense, BMU's output becomes, itself, an endogenous training signal.

The vector $\hat{v}_i^t \in \mathbb{R}^d$ defines a measure of unit relevance that can be used do determine which neurons are less significant for the representation of the input patterns. Such measure can be used, for instance, for deciding whether a neuron can be *silenced* or pruned from the network. CoRe neuron silencing resembles the unit commitment mechanism of the ART model [7]. In brief, an ART network consists of committed and uncommitted neurons: the latter set refers to *fresh* neurons that, initially, do not participate to the competition but that can be activated whenever none of the committed units represents enough well (with respect to a given *vigilance parameter*) the current input stimuli. Hence, an ART network starts with a pool of uncommitted neurons, activating them whenever a novel, surprising, stimuli in encountered; such enabling process continues until there are no more committable neurons. CoRe takes an alternative approach where, initially, all neurons are considered to be active and are allowed to participate to the competition. During learning their relevance factor is updated and they are silenced as soon as their $\hat{v}_i^t$ falls below a certain threshold of minimum relevance θ_{rel}. In practice, instead of silencing non relevant neurons, we simply prune them from the network, retaining only those units that are significant for coding the given data samples.

Following the same approach, the single components $\hat{v}_{il}^t$ of the relevance factor can be used to determine which prototype dimension c_{il} characterize better the input patterns assigned to the i-th neuron. The feature relevance can be used, on the one hand, to help profiling the data clusters discovered by the algorithm and, on the other hand, to prune irrelevant or noisy data components (see [3] for further details).

2.3 A Case Study in Breast Cancer Research

The learning algorithm described in the previous section can be readily applied to clustering tasks. In particular, each CoRe neuron u_i^o can be interpreted as a cluster detector with centroid c_i: starting with an initially large neural population, CoRe iteratively prunes irrelevant neurons until it converges to an estimate of the number of clusters in the data.

In this section, we study the performance of our neural approach on a recently published [4] dataset from breast cancer research. Information from biomarker proteins was collected from a consecutive series of 922 patients who underwent surgery for primary infiltrating breast cancer between 1983 and 1992 at the University of Ferrara [4]. This study analyzed proteins measured on preserved tissue samples, with exploratory aims and without diagnostic nor risk communication to the patient. Data on patient age, pathologic tumor size, histologic type, pathologic stage, and number of metastatic axillary lymph nodes were collected, as well as immunohistologic determinations of estrogen receptor (ER) status, progesterone receptors (PR) status, Ki-67/MIB-1 proliferation index (Pro), HER2/NEU (NEU), and P53 levels. Only the latter five biomarkers have been analyzed as part of this study: the 633 samples used in [4], for which is present complete information on all patho-biological measurements, are included in this work. Coherently with [4], the expression values of ER, PR, and NEU have been discretized to the following percentages: 0%, 10%, 25%, 50%, 75%, and 100%. Percentages of Ki-67- and p53-expressing cells were recorded without discretization.

CoRe is used to analyze the dataset and discover homogeneous groups of samples, possibly sharing a common bio-medical trait. Its results are compared with those obtained by three model-based algorithms, that are

- Gaussian Mixture Model (GMM) [9]: fits input samples to a mixture of Gaussians using the Expectation Maximization (EM) algorithm; the number of components (each identifying a single cluster) is also estimated from the data by pruning those mixtures with zero mixing weights.
- Variational Bayesian Gaussian Mixture (VBG) [10]: uses a variational method to fit a fully Bayesian model to the samples; as with GMM it determines the number of mixtures by component pruning.
- Variational Bayesian Mixtures with Splitting (VBS) [11]: similarly to VBM uses variational methods to fit the data to the model; the number of mixtures is determined by recursively splitting and pruning the existing components.

Additionally, the results obtained by each algorithm are compared with the sample labeling discovered in previous work [4] by Partition Around Medoids (PAM) and k-means (KM). The results are based on 50 runs of each algorithm, with random initial prototype positions. Both GMM and VBM are initialized with 20 mixtures, VBS is initialized with one component and CoRe starts with 30 units (for CoRe meta-parameter settings refer to [3]).

First, we focus on determining the most likely number K of sample groups in the data. Figure 2.2 shows the histogram of the cluster numbers estimated by the four algorithms during the 50 runs. The distribution of K for GMM and VBG closely resembles a Gaussian centered on 7 and 5, respectively. CoRe and VBS produce sharper hypothesis: the former, in particular, suggests that the data can be grouped in 4 or 5 clusters, with few runs terminating with 6 clusters. The behavior of VBS

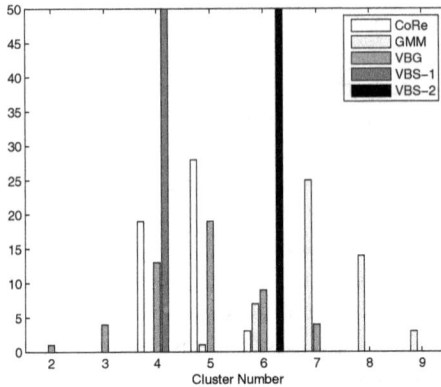

Fig. 2.2 Distribution of the cluster number estimate for the 50 independent runs of the algorithms

is peculiar: this algorithm behaves similarly to a wrapper approach that tentatively splits the existing components and locally tests the existence of additional clusters, eventually pruning the newly generated mixture; this process is iterated until the splitting test fails for all the components. Within this stopping criterion, VBS always converged to a solution where $K = 6$ (see the VBS-2 bar in Fig. 2.2). However, by taking a closer look at VBS learning dynamics one discovers that the maximum likelihood solution is obtained always for $K = 4$ (that is the VBS-1 solution in Fig. 2.2), while $K = 5$ is the second best-scoring solution with respect to the likelihood.

These results suggest that the most likely cluster number estimates are $K = 4$, $K = 5$ and $K = 6$. To evaluate the agreement between the data partitions produced by the different algorithms, we compared them with a baseline k-means clustering and with the sample labels discovered by PAM in [4]. Table 1 summarizes the concordance of the generated solutions in terms of the κ statistics [12]: the GMM model is not shown since it produced results only for $K = 6$, with a minimal overlap between its solutions and those produced by the other algorithms. Besides GMM, the model based algorithms seem to produce partitions that are quite uncorrelated with respect to those generated by the other algorithms. In particular, VBG produces groupings that are only marginally supported both by CoRe and the hard clustering algorithms, as well as by VBS. This latter model shows a fair agreement with CoRe on the hypotheses $K = 4$ and $K = 6$ and a rather weaker concordance on $K = 5$.

The κ values in Table 8.1 show a substantial agreement between CoRe and k-means, especially for the hypotheses K=4 and K=5, that are those most strongly advised by CoRe. The concordance with the PAM labels is analyzed only on the $K = 4$ hypothesis since this is the cluster number estimated by [4] using PAM with model selection indices (e.g. Gap statistics, KL, etc.). The KM solution shows the highest agreement with PAM labels, confirming the results in [4], while Gaussian-based algorithms seem to find different solutions with respect to both KM and PAM. As a general comment, the model-based algorithms seem unable to converge to a shared

K	CoRe				VBG			VBS		PAM
	KM	PAM	VBS	VBG	KM	PAM	VBS	KM	PAM	KM
$K=4$	0.83	0.83	0.65	0.46	0.54	0.47	0.36	0.76	0.74	0.96
$K=5$	0.83	X	0.54	0.38	0.36	X	0.53	0.65	X	X
$K=6$	0.68	X	0.61	X	0.23	X	0.52	0.49	X	X

Table 2.1 Clustering concordance evaluated by κ statistics: X's indicates when model comparison was not possible.

sample classification, producing quite discording dataset partitions. CoRe, on the other hand, produces solutions that are a trade-off between the concordant sample classification produced by PAM and KM, and the alternative partition discovered by VBS.

To gather a better insight into the CoRe's cluster profiles we looked at the relevance $\hat{v}_{il}$ of the biomarkers characterizing CoRe's sample groups. Figure 2.3.a and 2.3.b show CoRe's feature relevance for the K=4 and K=5 models, respectively. The relevance plot suggests that the proliferation marker PRO does not play a key role in determining the group membership of the samples. In both hypotheses, there appears to be a single cluster that is strongly characterized by the NEU biomarker and that corresponds to the most aggressive tumor profile [4]. The P53 biomarker appears to be strongly discriminant for the single C_1^4 cluster (in the $K=4$ scenario). On the other hand, in the $K=5$ hypothesis, the P53 covariate becomes relevant for two clusters, that are C_4^5 (corresponding to C_1^4) and C_3^5. Finally, the ER and PR markers seem to follow a correlated relevance pattern: in other words, for a given cluster, they are either both relevant or both irrelevant in determining the cluster membership.

Table 2.2 shows a detailed comparison between PAM and CoRe results in terms of their confusion matrices: when $K=4$ both algorithms seem to agree on the existence of a large C_2^4-P_1 group, that CoRe characterizes with an high PR-relevance and a medium relevance of the ER marker (see Fig.2.3.a). This hypothesis seems to be consistent with the distribution of the PR and ER values in the Core and PAM clusters (see Fig.2.4 and Fig.2.5, respectively). In particular, the boxplots in Fig.2.4 and Fig.2.5 show that the C_2^4-P_1 group is the sole characterized by high PR values: therefore CoRe has correctly identified PR as the most discriminative marker for the cluster. On the other hand, the ER marker shows high levels both for C_2^4-P_1 and C_3^4-P_2 obtaining a lower relevance due to its smaller discriminative effect.

The results in Table 2.2 show that part of PAM's P_1 cluster is split by CoRe and assigned to C_3^4, that roughly corresponds to PAM's P_2: Fig.2.4.c and Fig.2.5.b confirm that the two clusters have a similar markers' distribution. Notice that C_3^4 has a very similar pattern of feature relevance with respect to C_2^4, except for a slightly higher significance of the ER covariate (see Fig.2.3.a). In particular, the high PR relevance is due to the fact that these two clusters have a very similar markers' distribution except for the PR covariate, that characterizes C_2^4 and C_3^4 respectively by a high and low marker expression (see the boxplots in Fig.2.4.b and Fig.2.4.c).

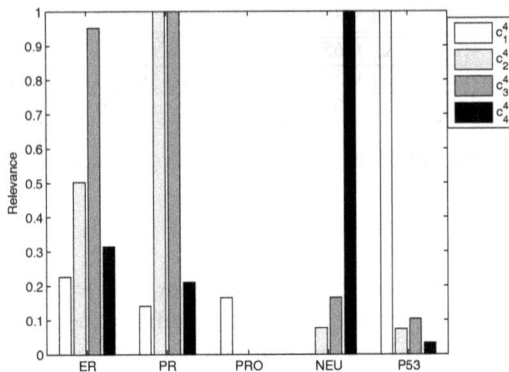

(a)

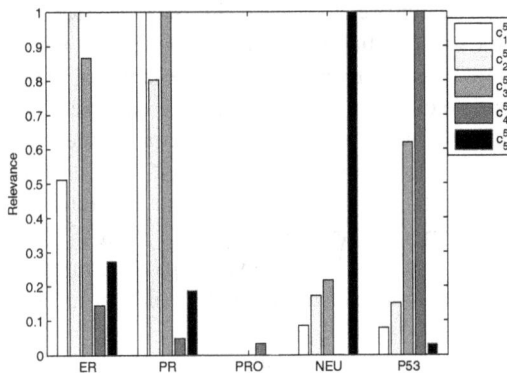

(b)

Fig. 2.3 CoRe relevance factors of the 5 features for the (a) 4 clusters and (b) 5 clusters scenario

	CoRe ($K=4$)				CoRe ($K=5$)				
PAM	C_1^4	C_2^4	C_3^4	C_4^4	C_1^5	C_2^5	C_3^5	C_4^5	C_5^5
P_1	9	210	32	5	193	28	32	3	0
P_2	24	0	159	24	0	76	131	0	0
P_3	76	0	0	15	0	0	1	79	11
P_4	0	0	3	76	0	4	5	0	70

Table 2.2 Samples cross-distribution between PAM and CoRe ($K = 4$ and $K = 5$)

The two smaller groups identified by PAM, i.e. P_3 and P_4, roughly correspond to CoRe's C_1^4 and C_4^4 groups, respectively. CoRe clusters, however, are larger than PAM's P_3 and P_4, since they gather samples from the highly populated P_2 group (see the confusion matrix in Table 2.2). Interestingly, CoRe isolates sharply the small P_3

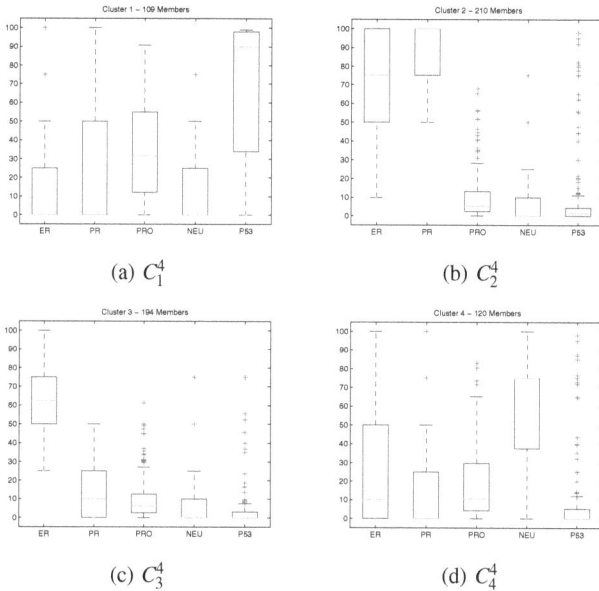

(a) C_1^4

(b) C_2^4

(c) C_3^4

(d) C_4^4

Fig. 2.4 Boxplots of CoRe cluster profiles for the $K = 4$ solution

and P_4 groups in the $K = 5$ hypothesis: the results in Table 2 show a strong P_3-C_4^5 and P_4-C_5^5 overlap. Such an overlap is confirmed by the similarity of the corresponding marker distributions in Fig.2.5 and Fig.2.6. Seemingly, the addition of the cluster C_2^5 attracted those *spurious* P_2 samples that, in the $K = 4$ hypothesis, were assigned by CoRe to partition C_1^4 and C_4^4. In our opinion, this suggests the existence of a fifth cluster that is situated *in between* the P_2, P_3 and P_4 groups from PAM's solution. Already in [4], the authors pointed out the heterogeneity of the samples in P_2 with respect to the other three clusters, hypothesizing the existence of a further subdivision of P_2 in smaller subgroups. The outcome of our analysis seems to confirm this initial hypothesis. In particular, by analyzing the biomedical features associated with the clusters we can derive interesting prognostic consequences underlying the existence of a fifth tumor subgroup.

The results in [4] shows that, in general, the P_1 and P_2 samples are associated with the least aggressive tumor subgroups and are characterized by minimal-change lesions as well as hormone sensitivity. On the other hand, individuals from the P_3 and P_4 groups tend to develop an increased number of metastatic lymph nodes and the corresponding tumors are characterized by an higher proliferative rate and by more frequent oncogene suppressor alterations. More in detail, the analysis in [4] shows that the P_4 group, characterized by high levels of HER2/NEU (see Fig.2.5.d), has an intermediate prognosis amongst patients non treated with hormone therapy but has the poorest response among the treated patients. Conversely, the P_3 group shows the worse event free survival (EFS) rate among the non-treated patients. The

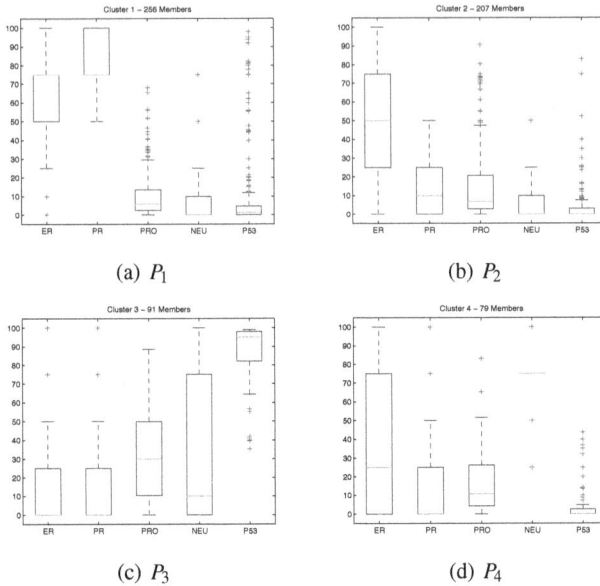

(a) P_1

(b) P_2

(c) P_3

(d) P_4

Fig. 2.5 Boxplots of PAM cluster profiles

Kaplan-Meier curves of cluster P_1 and P_2 for non-treated cases essentially overlaps (see Figure 5 in [4]), showing an event free survival rate above 80% for a five-years period, thus confirming the low aggressiveness of these two tumor subgroups compared to P_3 and P_4. However, it is worth noticing a peculiar behavior in the Kaplan-Meier curve of the P_2 cluster for patients treated with hormone therapy: while the P_1 group shows an EFS pattern that is close to the non-treated cases, P_2 shows a notably different behavior after the first 30 months. In fact, the event free survival of P_2 patients drops to values close to those obtained by individuals in the P_3 group. This behavior might confirm the existence of two tumor subtypes inside the original P_2 group: our hypothesis is that the CoRe clusters C_2^5 and C_3^5 correspond to these two tumor subtypes. In particular, the former group might be related to individuals that are responsive to hormone therapy, while the latter can possibly describe a tumor bioprofile characterized by a stronger resistance to hormone treatment.

Our conjecture is supported by the analysis of the marker profiles of the C_2^5 and C_3^5 groups (see Fig.2.6.b and Fig.2.6.c,respectively). The C_2^5 cluster is characterized by an high expression of the estrogen receptor (ER), which is considered a major predictor of response to hormone therapy [4]. In addition, standing to CoRe's relevance measure in Fig.2.3.b, the ER covariate is the most significant feature in determining the membership of samples to cluster C_2^5. The C_3^5 group, on the other hand, is characterized by a low activity of both the ER and PR hormone receptors (see Fig.2.6.c): in our opinion, this aspect might have triggered the reduced response to treatment of some P_2 individuals. From the boxplots in Fig.2.6.c and Fig.2.6.d it is clear that the key difference between the profiles of cluster C_3^5 and C_4^5-P_3 lies in

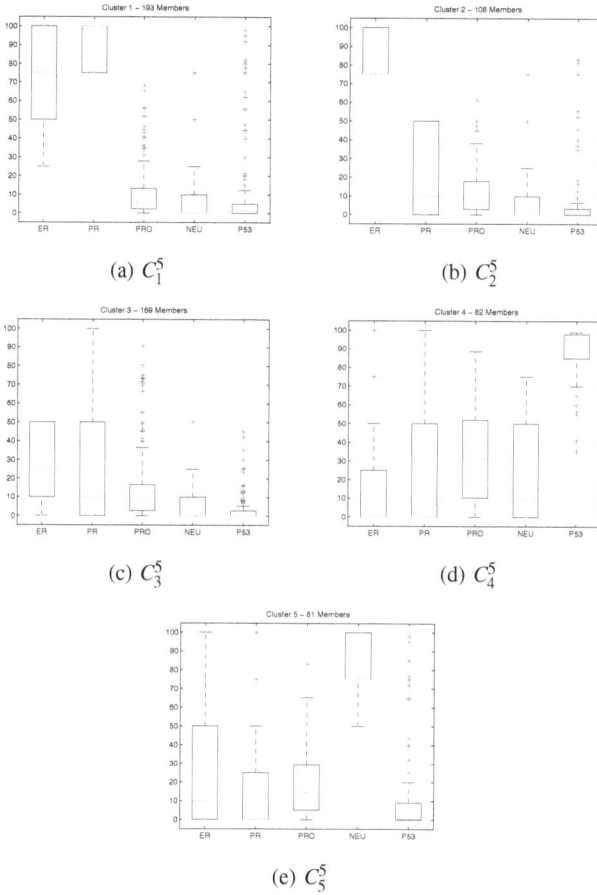

(a) C_1^5

(b) C_2^5

(c) C_3^5

(d) C_4^5

(e) C_5^5

Fig. 2.6 Boxplots of CoRe cluster profiles for the $K = 5$ solution

the expression of the P53 marker (which is confirmed by its high CoRe relevance in both clusters). In a sense, the identification of the C_3^5 cluster suggests the existence of an intermediate tumor subgroup that situates between the aggressive cancer form of the C_4^5-P_3 individuals and the (possibly) treatment-responsive cases in C_2^5.

2.4 Conclusion

We have studied unsupervised cluster number estimation on a breast cancer dataset, comparing the performance of hard clustering, neural and model based algorithms. The experimental results showed how the CoRe algorithm can be used to unsupervisedly explore biomedical data, estimating the number of the clusters in the dataset and producing a measure of the relevance of the samples' covariates that can be used to identify significant markers of the discovered bio-profiles. The comparison with the Bayesian models has highlighted how their performance can be seriously affected by the nature of the dataset, preventing them to reliably estimate the data clusters.

The results produced by CoRe seem to confirm the hypothesis presented in [4] concerning the existence of a fifth tumor subgroup in the breast cancer case-study. Our hypothesis, that shall by validated by clinical studies, is that such sub-group might differentiate two breast cancer subgroups. The first form is characterized by a high expression of the estrogen receptors which should determine a low aggressiveness and a good response to hormonal therapy. The latter sub-group should differentiate a more aggressive tumoral type that is less responsive to treatment and is characterized by an event free survival profile close to that of individuals with high P53 expression.

Acknowledgements

This research was performed with the financial support of the Italian Ministry of University and Research (MIUR), PRIN prot. 2005030003-005. This work was carried out with partial funding from the Biopattern Network of Excellence FP6/2002/IST/1; N IST-2002-508803 (http://www.biopattern.org).

References

1. Vellido, A., Lisboa, P.J.G.: Neural networks and other machine learning methods in cancer research. In: Proc. of the 9th International Work-Conference on Artificial Neural Networks (IWANN'07). Volume 4507 of Lecture Notes in Computer Science., Springer (2007) 964–971
2. Bacciu, D., Starita, A.: A robust bio-inspired clustering algorithm for the automatic determination of unknown cluster number. In: Proceeding of the 2007 International Joint Conference on Neural Networks (IJCNN'07), IEEE (2007) 1314 – 1319
3. Bacciu, D., Micheli, A., Starita, A.: Simultaneous clustering and feature ranking by competitive repetition suppression learning with application to gene data analysis. In: Proceeding of the 2007 Conference on Computational Intelligence in Medicine and Healthcare (CIMED'07). (2007)
4. Ambrogi, F., Biganzoli, E., Querzoli, P., Ferretti, S., Boracchi, P., Alberti, S., Marubini, E., Nenci, I.: Molecular subtyping of breast cancer from traditional tumor marker profiles using parallel clustering methods. Clin Cancer Res. **12**(1) (2006) 781–790
5. Kohonen, T.: Self-organized formation of topologically correct feature maps. Biol Cybern **43**(1) (1982) 59 – 69
6. Fritzke, B.: A growing neural gas network learns topologies. In Tesauro, G., Touretzky, D.S., Leen, T.K., eds.: Advances in Neural Information Processing Systems 7. MIT Press, (Cambridge MA) 625–632
7. Carpenter, G., Grossberg, S.: The ART of adaptive pattern recognition by a self-organizing neural network. Computer **21**(3) (1988) 77–88

8. Yair, E., Zeger, K., Gersho, A.: Competitive learning and soft competition for vector quantizer design. IEEE T SIGNAL PROCES **40**(2) (Feb 1992) 294–309
9. Figueiredo, M.A.T., Jain, A.K.: Unsupervised learning of finite mixture models. IEEE T PATTERN ANAL **24**(3) (2002) 381–396
10. Corduneanu, A., Bishop, C.: Variational bayesian model selection for mixture distributions. In Richardson, T., Jaakkola, T., eds.: Artificial Intelligence and Statistics, Morgan Kaufmann (2001) 27–34
11. Constantinopoulos, C., Likas, A.: Unsupervised learning of gaussian mixtures based on variational component splitting. IEEE T NEURAL NETWOR **18**(3) (2007) 745–755
12. Cohen, J.: A coefficient of agreement for nominal scales. Educational and Psychological Measurement **20** (1960) 37–46

Chapter 3
Downsizing Multigenic Predictors of the Response to Preoperative Chemotherapy in Breast Cancer

René Natowicz, Roberto Incitti, Roman Rouzier, Arben Çela, Antônio Braga, Euler Horta, Thiago Rodrigues, Marcelo Costa, and Carmen D. M. Pataro

Abstract We present a method for designing efficient multigenic predictors with few probes and its application to the prediction of the response to preoperative chemotherapy in breast cancer. In this study, each DNA probe was regarded as an elementary predictor of the response to the chemotherapy and the probes which were selected performed a faithful sampling of the training dataset. In a first stage of the study, the prediction delivered by a multigenic predictor was that of the majority of the elementary predictions of its probes. For the data set at hand, the best majority decision predictor (MD predictor) had 30 probes. It significantly outperformed the best predictor previously published, which was designed on probes that had been selected by p-value of a t-test. In a second stage, the majority decision was replaced by a support vector machine (SVM) with linear kernel. With the same set of probes, the performances of the SVM predictor were slightly better for both training and testing sets of data. The main improvement was that the performances of the best MD predictor were achieved by the SVM predictors with only 17 probes. This more than 40% downsizing of the predictors is an interesting property for the potential use of the predictors in clinical routine, and for the task of modeling the biological mechanisms underlying the patient's response to the chemotherapy.

3.1 Introduction

Nowadays, adjuvant and neoadjuvant (preoperative) administration of chemotherapy is based on prognostic factors, not on predictive ones. It is well known that the prognostic factors do not provide enough information for tailoring the treatment to the individual patients. Hence, nearly all breast cancer patients are given a stan-

René Natowicz
Université Paris-Est, ESIEE-Paris, France, e-mail: `r.natowicz@esiee.fr`

dard chemotherapy treatment, despite their potentially poor response to the therapy, adverse side effects, and healthcare costs.

The ability to predict the patients' response to the chemotherapy would be of high interest in the treatment of breast cancer for avoiding useless chemotherapy treatments and for selecting the most effective regimen for every patient. To this end, no single factor or biomarker ever has been in position to discriminate the patients who would respond to the treatment from those who would not. It appears that primary chemotherapy provides an ideal opportunity to correlate the gene expressions with the response to the treatment. Although gene expression microarrays provide novel tools and hold great promise in cancer research, the achievements in terms of improved prediction of drug sensitivity have been thus far rather moderate [1]. A strategy for translating microarray profiles into efficient clinical tests could consist in identifying small diagnostic gene-expression profiles with the help of microarrays then, in a second step, to validate the clinical usefulness of these genes, either retrospectively or prospectively, by making use of a simple and robust conventional assay, such as the quantitative reverse-transcriptase polymerase chain reaction (RT-PCR). Such a strategy requires microarrays analysis methods able to provide oncogenic signatures made out of few probe sets.

In the present study, every selected probe delivered an elementary prediction of the response to the treatment: pathologic complete response (*pcr*), residual disease (*nopcr*), or *unspecified*. In a first stage, we have defined very simple predictors whose predictions of the patient's response were those of the majority of their probes' elementary predictions: *PCR* if the majority of the probes predicted the response to be *pcr*; *NoPCR* if the majority was *nopcr*; and *UNSPECIFIED* in case of tie. In the second part of the study, the classification criterion of majority decision (MD) was replaced by support vector machines (SVM) with linear kernel. The resulting classifiers had slightly better training and testing performances with the same numbers of probes. Moreover, the performances of the best MD predictor were achieved by a SVM predictor with significantly less probes, 17 probes instead of 30, i.e. were more than 40% downsized.

In this paper, we will present the low level treatment through which a probe delivered an elementary prediction of the patient's response; the valuation function by which the probes were ranked then selected in this ranking; and we will give the performances of the MD predictors then those of the SMV predictors for the same dataset.

3.2 Patients and Data

This work was conducted based on data from Hess *et al*. The clinical trial was conducted at the Nellie B. Connally Breast Center of The University of Texas M.D. Anderson Cancer Center [2]. One hundred thirty-three patients with stage I-III breast cancer were included. All patients underwent a single-pass, pretreatment fine-needle aspiration of the primary breast tumor before starting chemotherapy. Pretreatment gene expression profiling was performed with oligonucleotide microarrays (Affymetrix U133A) on fine-needle aspiration specimens. Patient cases were separated into patient *training cases* (82 cases) and patient *testing cases* (51 patient cases). At the completion of neoadjuvant chemotherapy, all patients had surgical resection of the tumor bed, with negative margins. Pathologic complete response (PCR) was defined as no histopathologic evidence of any residual invasive cancer cells in the breast, whereas residual disease was defined as any residual cancer cells after histopathologic study. The low level treatment of the microarray data was performed by software dCHIP V1.3 to generate probe level intensities. This program normalizes all arrays to one standard array that represents a chip with median overall intensity. Finally, normalized gene expression values were transformed to the $\log_{10}$ scale for analysis because in microarrays, the log-values of the expression levels are closer to normal distributions than the non-transformed ones.

The set of training cases was composed of 82 patient data, each of which being the response to the treatment and the expression levels of the 22283 DNA probes. Among the training set, the response to the treatment was PCR for 21 patient cases and NoPCR for 61 cases. The testing set was composed of 51 patient data among which the response to the treatment was PCR for 13 patient cases and NoPCR for 38 patients. Hence, the ratios of PCR to NoPCR patient cases were the same for both the training and testing datasets.

3.3 Probes Valuation

A first research on these data is reported In [2]. In this study, the authors the probes were selected in the ranking of the p-value of a t-test, and the study was more focused on the question of choosing the best decision model for the prediction. To this end, the authors checked several models of classification: k-nearest neighbors, support vector machine, diagonal linear discriminant analysis, with various parameters for each of them. A total of 780 different classifiers have been evaluated and statistically assessed, all of them taking as input the expression levels of the probes selected by the p-values of a t-test. This study has shown that, for this criterion of probes selection, the best predictor was a one taking as input the expression levels of the 30 probes of least p-values, these expression levels being weighted thanks to a diagonal linear discriminant analysis (predictor DLDA-30). For the set of validation cases, the performances of the DLDA-30 predictor were[1]: accuracy=0.76, sensitivity=0.92, specificity=0.71, PPV=0.52, NPV=0.96.

---

[1] Definitions: let TP and TN be the numbers of true positives and negatives returned by a given predictor, and let FP and FN be the numbers of false positives and negatives. The accuracy of the predictor is the proportion of correctly predicted cases, the sensitivity that of correctly predicted PCR cases, the specificity is the proportion of correctly predicted NoPCR cases, the positive predictive value (PPV) is the probability of a case predicted as PCR to be a PCR case, and the negative predictive value (NPV) is the probability of a case predicted NoPCR to be a NoPCR case. Therefore, these criteria are: accuracy=(TP+TN)/(TP+FN+TN+FP), sensitivity=(TP)/(TP+FN), speci-

Hence, the predictor DLDA-30 had very high sensitivity and negative predictive values. Taking the results of this study as a starting point, we wanted to design predictors with higher specificity values, while preserving the very high sensitivity value of the DLDA-30 predictor. Furthermore, because the study reported in [2] had put the emphasis on the classification models and was very complete in this regard, we have decided to investigate the process of probes selection, with the wish of improving the performances with models of classification as simple as possible.

For the dataset at hand, it has appeared to us that the probes selection criterion of p-value to a t-test favored probes that mainly gave some information on the membership of the patients to the NoPCR class. For instance, the probe of smallest p-value was the probe 203929_s_at, of the gene MAPT. The box-plot of its expression levels (fig. 3.1) shows that given a high expression level, the probability of NoPCR membership was high. It also shows that a low expression level for this probe did not provide any information on the patient's class: in this case, the probability to predict a PCR was more or less that of a random choice with probabilities $\frac{P}{P+N}$ and $\frac{N}{P+N}$, where P and N are the respective numbers of PCR and NoPCR cases of the training set.

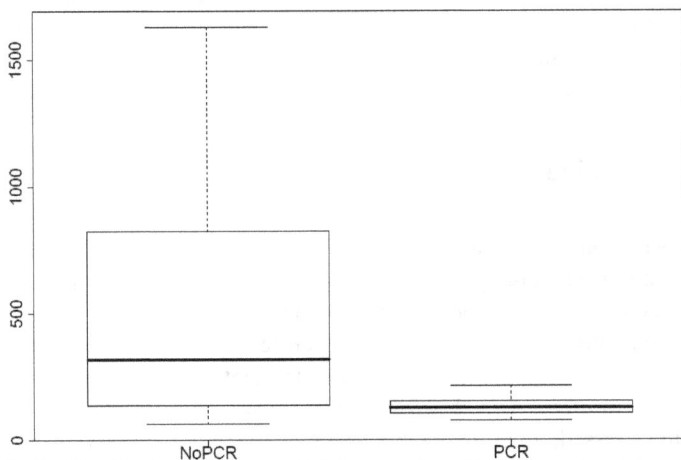

Fig. 3.1 Probe 203929_s_at of the gene MAPT, probe of smallest p-value to a t-test: box-plot of the expression levels.

Because we aimed at designing predictors with better performances and simple classification criteria, we were interested in putting into light DNA probes conveying information on the two classes of patients, PCR and NoPCR. Qualitatively speaking, we wanted to find probes whose interval of expression levels respectively computed on the PCR and on the NoPCR subsets of learning cases, would have a "small intersection", and which would deliver an information non biased by the overrepresentation of the NoPCR cases in the learning set.

---

ficity=(TN)/(TN+FP), the positive predictive value PPV is (TP)/(TP+FP) and the negative predictive value is (TN)/(TN+FN).

To this end, we have chosen to assign two sets of expression levels to each probe s, the sets $E_p(s)$ and $E_n(s)$, computed from the training data as follows [3, 4]. Let $m_p(s)$ and $sd_p(s)$ be the mean and standard deviation of the expression levels of the probe s for the PCR training cases, and let $m_n(s)$ and $sd_n(s)$ be those of the NoPCR training cases. The set of expression levels of the PCR training cases was defined as the set difference $E_p(s)$, $E_p(s) = [m_p(s) - sd_p(s), m_p(s) + sd_p(s)] \setminus [m_n(s) - sd_n(s), m_n(s) + sd_n(s)]$ and conversely for the NoPCR training cases, $E_n(s) = [m_n(s) - sd_n(s), m_n(s) + sd_n(s)] \setminus [m_p(s) - sd_p(s), m_p(s) + sd_p(s)]$.

Discrete probes' predictions. For any patient case, the individual prediction of a probe was a discrete value in the set $\{pcr, nopcr, unspecified\}$: *pcr* if the expression level of patient p was in the interval $E_p(s)$ and *nopcr* if it was in $E_n(s)$. Otherwise, the individual prediction value was *unspecified*.

Probes' valuation function. Let $p(s)$ be the number of PCR training cases correctly predicted *pcr* by the probe s, and let $n(s)$ be the number of NoPCR training cases correctly predicted *nopcr* by the probe. The valuation function of the probes was defined so as to favor probes which correctly predicted high numbers of training cases and whose sets of correctly predicted training cases were 'good' samplings of the training set. To this end, we have considered the sensitivity and specificity values of the probe s, i.e. the ratios $p(s)/P$ and $n(s)/N$ of correctly predicted training cases. The valuation function $v(s)$, $v(s) \in [0, 1]$, was defined as the sum of its sensitivity and specificity, $v(s) = 0.5 \times \left(\frac{p(s)}{P} + \frac{n(s)}{N} \right)$ (the coefficient ensures that the valuation is in the unit interval).

The figure 3.2 is the box-plot of the expression levels of a DNA probe of the gene BTG3, for the patients of the training set. This probe was one of the two equally top ranked probes, (cf. table 3.3.)

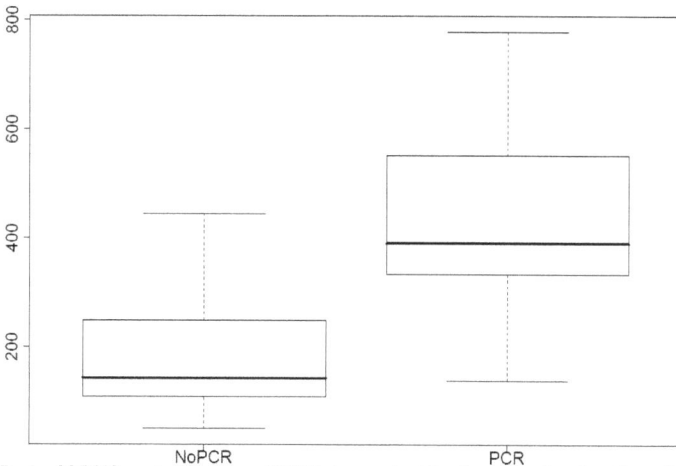

Fig. 3.2 Probe 205548_s_at of the gene BTG3, top ranked for the valuation function $v(s)$: box-plot of the expression levels.

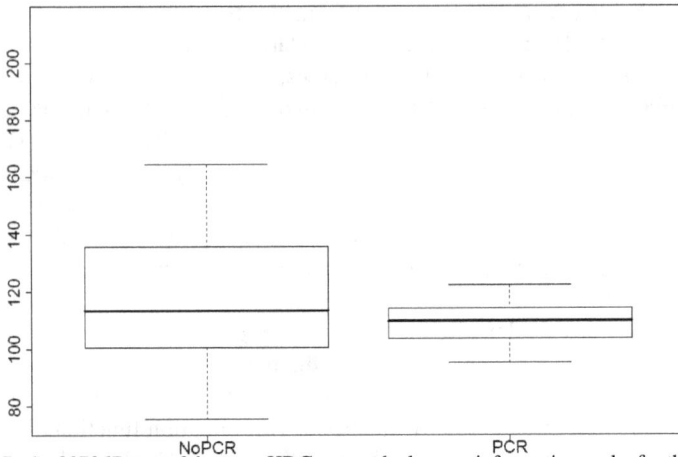

Fig. 3.3 Probe 207067_s_at of the gene HDC, top ranked mono-informative probe for the valuation function $v(s)$: box-plot of the expression levels.

From this figure, one can see that, given that the expression level of this probe was high, there was a high probability for the patient to have a pathologic complete response and symmetrically, given that the expression level was low, a high probability of residual disease. Hence, this probe delivered an information about both PCR and NoPCR classes membership. Up to a high rank, the probes selected by decreasing values conveyed information on both classes. In the ranking of the valuation function, the first *mono-informative* probe (giving information about only one class of patients) was at rank 63 (probe 207067_s_at of the gene HDC, cf. fig. 3.3.)

3.4 Multigenic Predictors with Majority Decision

We have defined the k-probes majority decision predictor (MD predictor) as the k top ranked probes for the valuation function $v(s)$ together with the classification criterion of majority decision: for each patient case, the prediction was PCR if the number of elementary *pcr* predictions was strictly greater than the number of elementary *nopcr* ones, the prediction was NoPCR for the converse situation, and UNSPECIFIED in case of tie. In figure 3.4 are the training and testing accuracies of the first 41 MD k-predictors ($0 \leq k \leq 40$.) In table 3.3 are the numbers of false positives and false negatives of the same set of MD k-predictors.

The predictor of highest training accuracy was the 38-probes predictor: acc.=0.85, sens.=0.86, specif.=0.85, npv=0.945 (negative predictive value), ppv=0.67 (positive predictive value), corresponding to 3 FN (false negatives) and 9 FP (false positives) out of 21 PCR and 61 NoPCR cases. Its performances on the test set were:

Gene	Probe	$v(s)$	$p(s)$	$n(s)$	Gene	Probe	$v(s)$	$p(s)$	$n(s)$
BTG3	213134_x_at	0.61	12	40	GATA3	209602_s_at	0.41	13	13
BTG3	205548_s_at	0.61	12	40	BBS4	212745_s_at	0.41	3	42
GATA3	209604_s_at	0.59	15	29	DAPK1	203139_at	0.41	9	24
GATA3	209603_at	0.49	12	26	SAS	203226_s_at	0.40	7	29
THRAP2	212207_at	0.46	8	34	FLJ10916	219044_at	0.40	8	26
SCCPDH	201826_s_at	0.46	12	22	E2F3	203693_s_at	0.40	8	26
SIL	205339_at	0.45	10	27	AHNAK	220016_at	0.40	9	23
KRT7	209016_s_at	0.45	6	38	KLHDC3	214383_x_at	0.40	9	23
MCM5	201755_at	0.45	7	35	SFRS12	212721_at	0.40	9	23
NME3	204862_s_at	0.44	10	25	SRPK1	202200_s_at	0.39	6	31
METRN	219051_x_at	0.44	11	22	CXCR4	217028_at	0.39	8	25
PDE4B	211302_s_at	0.43	9	27	KIF3A	213623_at	0.39	8	25
PHF15	212660_at	0.42	7	32	MGC4771	210723_x_at	0.39	8	25
SSR1	200891_s_at	0.42	7	32	C11orf15	218065_s_at	0.39	9	22
PISD	202392_s_at	0.42	11	20	CELSR1	41660_at	0.39	12	13
MELK	204825_at	0.41	8	28	LAD1	203287_at	0.39	4	36
CA12	215867_x_at	0.41	10	22	LU	203009_at	0.38	6	30
CA12	214164_x_at	0.41	10	22	LIPE	213855_s_at	0.38	7	27
MAPK3	212046_x_at	0.41	10	22	GAMT	205354_at	0.38	7	27

Table 3.1 The 38 probes of highest valuations. Gene: gene name in Hugo Gene nomenclature[5]; probe: reference of the Affymetrix DNA probe set; $v(s)$: probe valuation; $p(s)$, $n(s)$: numbers of correct pcr and nopcr predictions for the 21 PCR and 61 NoPCR cases of the training set. Total numbers of pcr and nopcr predictions: 301 and 900, ratio=0.33.

Gene	Probe	$v(s)$	rank	$p(s)$	$n(s)$	Gene	Probe	$v(s)$	rank	$p(s)$	$n(s)$
MAPT	203929_s_at	0.22	780	0	28	AMFR	202204_s_at	0.23	662	0	29
MAPT	203930_s_at	0.291	218	2	30	CTNND2	209617_s_at	0.27	337	0	33
BBS4	212745_s_at	0.41	21	3	42	GAMT	205354_at	0.38	38	7	27
MAPT	203928_x_at	0.22	781	0	28	CA12	204509_at	0.24	566	1	27
THRAP2	212207_at	0.46	5	8	34	FGFR1OP	214124_x_at	0.37	52	6	28
MBTPS1	217542_at	0.26	391	0	32	KIAA1467	213234_at	0.25	475	3	22
MAPT	206401_s_at	0.22	900	0	27	METRN	219051_x_at	0.44	11	11	22
PDGFRA	215304_at	0.32	118	4	28	FLJ10916	219044_at	0.40	24	8	26
ZNF552	219741_x_at	0.24	564	1	27	E2F3	203693_s_at	0.40	25	8	26
RAMP1	204916_at	0.22	774	0	28	ERBB4	214053_at	0.21	1040	0	26
BECN1	208945_s_at	0.30	165	4	26	JMJD2B	215616_s_at	0.37	45	7	26
BTG3	213134_x_at	0.61	1	12	40	RRM2	209773_s_at	0.37	51	3	37
SCUBE2	219197_s_at	0.15	3078	0	19	FLJ12650	219438_at	0.27	293	0	34
MELK	204825_at	0.41	16	8	28	GFRA1	205696_s_at	0.18	1994	0	22
BTG3	205548_s_at	0.61	2	12	40	IGFBP4	201508_at	0.38	39	7	27

Table 3.2 The 30 probes of smallest p-values to the t-test [1] (*ranks* are the probes' rankings for the valuation function $v(s)$.) Total numbers of pcr and no pcr predictions: 123 and 894, ratio=0.13

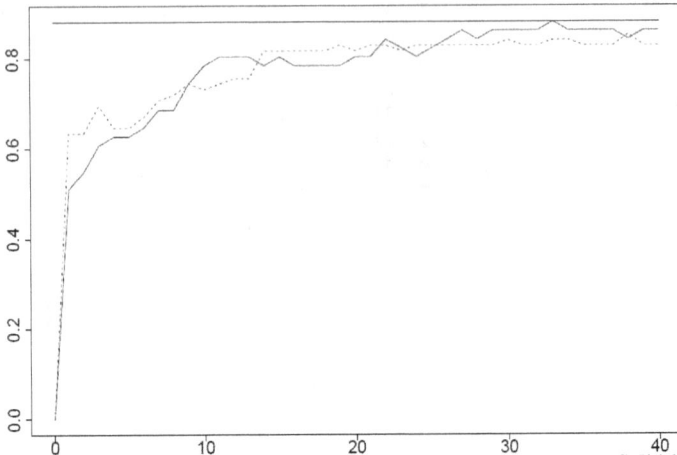

Fig. 3.4 Training and testing accuracies of the majority decision predictors. Solid line: testing accuracy, dashed line: training accuracy, horizontal line: maximum testing accuracy (0.88) for $k = 33$ probes. X axis: number of probes.

acc.=0.84, sens.=0.77, specif.=0.87, npv=0.917, ppv=0.67, corresponding to 3 FN and 5 FP out of 13 PCR and 38 NoPCR cases.

The MD predictors of second highest training accuracy had $k = 30$, $k = 33$ and $k = 34$ probes: acc.=0.84, sens.=0.81, specif.=0.85, npv=0.929, ppv=0.65 (4 FN and 9 FP). On the set of testing cases the performances of the 30-probes predictor were: acc.=086, sens.=0.92, specif.=0.84, npv=0.970, ppv=0.67, corresponding to 1 FN and 6 FP. The performances of the 33 and 34-probes predictors were approximatively the same on the set of testing cases (cf. table 3.3.)

k	Training FP	FN	Test FP	FN	k	Training FP	FN	Test FP	FN	k	Training FP	FN	Test FP	FN	k	Training FP	FN	Test FP	FN
1	21	9	18	7	11	15	6	7	3	21	11	3	8	2	31	10	4	5	2
2	21	9	18	5	12	14	6	7	3	22	11	3	6	2	32	10	4	5	2
3	19	6	16	4	13	15	5	7	3	23	11	4	7	2	**33**	**9**	**4**	**4**	**2**
4	22	7	15	4	14	11	4	8	3	24	10	4	7	3	**34**	**9**	**4**	**5**	**2**
5	22	7	15	4	15	12	3	7	3	25	10	4	7	2	35	10	4	5	2
6	21	6	15	3	16	12	3	8	3	26	10	4	6	2	36	10	4	5	2
7	18	6	13	3	17	12	3	8	3	27	10	4	6	1	37	10	4	5	2
8	17	6	12	4	18	12	3	8	3	28	10	4	5	3	**38**	**9**	**3**	**5**	**3**
9	15	6	9	4	19	11	3	9	2	29	10	4	6	1	39	10	4	5	2
10	16	6	7	4	20	12	3	8	2	**30**	**9**	**4**	**6**	**1**	40	10	4	5	2

Table 3.3 Confusion table for majority decision predictors. FP, FN: number of false positives and false negatives for training and testing cases. In bold: 1st and 2nd maximum training accuracies.

3.5 Training Set Sampling

Because the criterion the most widely used for selecting DNA probes in microarray studies for cancer research is the p-value of a t-test[1], and because the predictors designed with probes selected by our valuation function outperformed those designed with probes selected by the p-value[3], we were interested in finding a parameter which could account for the difference of performances. It has appeared that the quality of the sampling performed by a given set of probes could be this explicative parameter.

The ratio of the numbers of PCR to NoPCR training cases of the data set at hand was $\frac{P}{N} = \frac{21}{61} = 0.34$. For the valuation $v(s)$, this ratio was in excellent agreement with that of the total numbers of *pcr* to *nopcr* correct predictions of the k top ranked probes. For $20 \leq k \leq 50$ probes, the values of the ratios were between 0.33 and 0.34, and below 20 probes, the ratios were between 0.30 and 0.38. For the set of 30 top ranked probes, the mean number of correct predictions per probe was 35.16 and the ratio of the *pcr* to *nopcr* numbers of predictions was 0.34 (equal to the ratio of PCR to NoPCR numbers of cases in the training set).

For the p-value of a t-test, the set of 30 top ranked probes comprised 11 *mono-informative* probes (in the present case, their numbers of pcr predictions was null, cf. table 3.3). The mean number of correct predictions per probe was 37.67 (approximatively equal to that of the probes selected according to our probes valuation function), but the ratio of the *pcr* to *nopcr* numbers of predictions was far lesser, this ratio value being precisely equal to that of the whole set of probes (0.14). From this, one could see that, with the same mean number of predictions per probe, the quality of the sampling performed by a set of probes, measured by the ratio of the pcr to nopcr numbers of predictions, could explain the difference of performances between the two methods of probes selection. The 30 probes of least p-values to a t-test performed a sampling in which the *nopcr* predictions were over-represented while the 30 probes of highest values $v(s)$ performed a more faithful sampling of the training set, the numbers of *pcr* and *nopcr* predictions being in the ratio of the numbers of PCR to NoPCR testing cases.

3.6 Multigenic Predictors with Support Vector Machine

We have defined the k-probes support vector machine predictor (SVM predictor) as the k top ranked probes for the valuation function $v(s)$ together with a linear kernel SVM [6], [7]. The maximum training accuracy of the SVM k-predictors (table 3.4 and figure 3.5) was achieved for $k = 15$ probes: acc.=0.88, sens.=0.90, specif.=0.87, npv=0.96, ppv=0.70, corresponding to 2 false negatives and 8 false positives (out of 21 PCR and 61 NoPCR cases). The testing performances of the 15-probes predictor were: acc.=0.82, sens.=0.92, specif.=0.79, npv=0.968, ppv=0.6, corresponding to 1 false negative and 8 false positives (out of 13 PCR and 38 NoPCR cases).

Hence, the SVM predictor of highest training accuracy had slightly better training performances than the best MD predictor, and approximatively the same testing performances, with twice less probes (15 vs. 30 probes).

The second maximum training accuracy was achieved with $k = 17, 18, 19, 20$ probes and $k = 27, 28$ probes. On the plateau from 17 to 20 probes, the predictors had approximatively the performances of the $k = 15$ probes predictor for both training and testing accuracies. Because of this plateau, and because the 15-probes predictor appeared isolated, we have considered that 17-probes was a safer lower limit for downsizing the best MD predictor. On the small plateau of two predictors showing the second maximum training accuracy, $k = 27, 28$ probes, the testing performances were better than those of the best MD predictor but, obviously, without significant downsizing. The testing performances were: acc.=0.88, sens.=0.92, spec.=0.87, npv=0.971, ppv=0.71, corresponding to 1 false negative and 5 false positives. One noticed that this plateau of two predictors was the beginning of a longer plateau of predictors, from $k = 27$ to $k = 32$ probes, showing approximatively the same training performances and the same, or even better, testing performances.

Hence, from these results, one might say that the performances of the best MD predictor, the MD 30-probes predictor, were achieved by the SVM 17-probes predictor, i.e. with 43% less probes. In particular, their sensitivity values on the test set were equal (0.92) and their negative predictive values were almost equal (31/32=0.967 for the SVM 17-probes vs. 32/33=0.970 for the MD 30-probes predictor). In addition, the SVM 17-probes predictor showed slightly better training performances than the MD 30-probes predictor.

k	Training FP	FN	Testing FP	FN	SV	k	Training FP	FN	Testing FP	FN	SV
1	21	4	18	3	82	11	13	4	7	3	74
2	11	9	5	8	82	12	10	2	8	2	48
3	11	9	7	5	82	13	12	2	9	1	48
4	7	10	2	7	82	14	10	2	7	1	51
5	10	7	5	5	82	**15**	**8**	**2**	**8**	**1**	**54**
6	5	9	2	8	82	16	10	2	7	2	51
7	9	7	2	5	82	**17**	**9**	**2**	**7**	**1**	**51**
8	7	7	3	5	82	**18**	**9**	**2**	**7**	**1**	**52**
9	7	8	2	5	82	**19**	**9**	**2**	**8**	**1**	**51**
10	11	7	4	4	82	**20**	**9**	**2**	**8**	**1**	**50**

k	Training FP	FN	Testing FP	FN	SV	k	Training FP	FN	Testing FP	FN	SV
21	9	3	7	1	71	31	8	4	4	1	79
22	9	3	6	1	67	32	8	4	4	1	79
23	9	3	6	1	82	33	9	4	4	1	80
24	9	4	5	1	72	34	9	3	5	1	80
25	9	4	6	1	79	35	10	4	5	1	80
26	10	4	6	1	82	36	11	2	8	1	82
27	**7**	**4**	**5**	**1**	**79**	37	10	3	5	1	80
28	**8**	**3**	**5**	**1**	**70**	38	11	2	8	1	82
29	8	4	4	1	77	39	11	3	7	1	82
30	8	4	5	1	78	40	9	4	5	2	82

Table 3.4 Confusion table for support vector machine predictors. FP, FN: number of false positives and false negatives for training and testing cases; SV: number of support vectors. In bold: maximum training accuracies.

Conclusion

With our approach of features selection, the DNA probes were regarded as elementary predictors of the response to the chemotherapy. We have presented a valuation function through which the probes behaving as good samplers of the training set were assigned high values. Two classifier models were evaluated for these probes: the non weighted majority decision among the elementary predictions of the probes (MD-predictors), and a support vector machine with linear kernel (SVM-predictors). The probes making up the predictors were selected in the ranking of their values. The MD-predictor of highest accuracy for the training data was designed with the 30 probes of highest values. On the independant testing set, its accuracy, sensitivity, specificity, negative and positive predictive values were respectively 0.86, 0.92, 0.84, 0.97, and 0.67, which significantly outperformed the best predictors designed on probes selected by the p-values of a t-test (whose expression levels had been weighted by a diagonal linear discriminant analysis). We tend to think that our method of probes selection has revealed relevant genes because the significant improvement of the predictors' performances have been obtained by using a very simple decision criterion: the majority decision. In a second stage, using a support vector machine with linear kernel instead of the majority decision, the performances

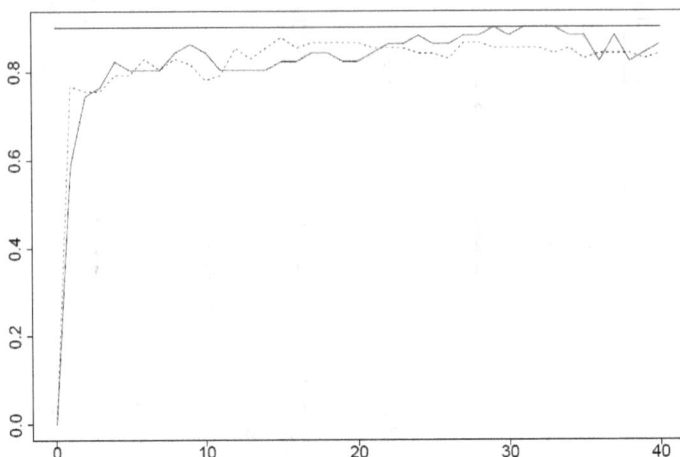

Fig. 3.5 Training and testing accuracies of the support vector machine with linear-kernel. Solid line: testing accuracy, dashed line: training accuracy, horizontal line : maximum testing accuracy (0.90) for $k = 29, 31, 32, 33$ probes. X axis: number of probes.

of the best MD predictor have been achieved with 17 probes instead of 30, hence the best MD predictor was downsized by more than 40%.

Because of the very high negative predictive value (0.97), such predictors could be of interest for supporting the decision of not allocating a patient to the treatment. Beside, the predictor would unadvisedly allocate 16% of the non-responder patients to the treatment (specificity=0.84). This result has to be compared to the nowadays almost systematic allocation to the treatment.

The small number of probes involved in the SVM predictor (17 probes) could allow the design of predictors at very low cost, which is an important issue for their potential use in clinical routine. Furthermore, because of the focusing on a small number of potentially relevant genes, this downsizing could also be an interesting property for modeling the biological mechanisms underlying the response to the chemotherapy.

Acknowledgments.

The authors would like to thank the CAPES-COFECUB french-brazilian cooperation program and the CNPq for their support.

References

1. Knudsen, S.: Cancer Diagnostics with DNA Microarrays. John Wiley & Sons, Inc. (2006)

2. Hess, K.R., Anderson, K., Symmans, W.F., Valero, V., Ibrahim N., Mejia, J.A., Booser, D., Theriault, R.L., Buzdar, A.U., Dempsey, P.J., Rouzier, R., Sneige, N., Ross, J.S., Vidaurre, T., Gómez, H.L., Hortobagyi, G.N., Pusztai, L.: Pharmacogenomic Predictor of Sensitivity to Preoperative Chemotherapy with Paclitaxel and Fluorouracil, Doxorubicin, and Cyclophosphamide in Breast Cancer. J. Clin. Onc. 24, pp. 4236-4244 (2006)
3. Natowicz, R., Incitti, R., Horta E.G., Charles, B., Guinot, P., Yan, K., Coutant, C., Andre, F., Pusztai, L., Rouzier, R.: Prediction of the Outcome of Preoperative Chemotherapy in Breast Cancer by DNA Probes that Convey Information on Both Complete and Non Complete Responses. BMC Bioinformatics 9, 149 (2008)
4. Natowicz, R., Braga, A.P., Incitti, R., Horta, E.G., Rouzier, R., Rodrigues, T.S, Costa, M.A.: A New Method of DNA Probes Selection and its Use with Multi-Objective Neural Networks for Predicting the Outcome of Breast Cancer Preoperative Chemotherapy. In: 16th European Symposium on Artificial Neural Networks, pp. 71-76. d-side ed., Evere, Belgium (2008)
5. Wain, M.H., Bruford, E.A., Lovering, R.C., Lush, M.J., Wright, M.W., Povey S.: Guidelines for Human Gene Nomenclature. Genomics, 79, pp. 464-470 (2002)
6. Cortes, C., Vapnik, V.: Support-vector networks. Machine Learning, 20, 273–297 (1995)
7. Burges, C.J.C.: A Tutorial on Support Vector Machines for Pattern Recognition. Data Mining and Knowledge Discovery, 2, pp. 121-167 (1998)

Chapter 4
Stratification of severity of illness indices and out-of-sample validation: a case study for breast cancer prognosis

Terence A. Etchells, Ana S. Fernandes, Ian H. Jarman, Josè M. Fonseca, and Paulo J.G. Lisboa

Abstract Prognostic modelling involves grouping patients by risk of adverse outcome, typically by stratifying a severity of illness index obtained from a classifier or survival model. The assignment of thresholds on the risk index depends of pairwise statistical significance tests, notably the log-rank test. This chapter proposes a new methodology to substantially improve the robustness of the stratification algorithm, by reference to a statistical and neural network prognostic study of longitudinal data from patients with operable breast cancer. This new methodology performance is then evaluated with an application to a new cohort form the same centre.

4.1 Introduction

Stratification of patients by risk of adverse outcome is central to clinical practice. This begins with modelling empirical data either with a classifier or a failure time model, depending on whether the data represent a snapshot in time of the patient's condition at diagnosis, or evolution of the disease over time in a longitudinal cohort study. Either way, the equivalent of the linear argument $\beta.x$ in a Generalised Linear Model defines a prognostic index that ranks patient data by severity of the illness. In the case of breast cancer, typically a piecewise linear model is used [1] from which the prognostic index can be derived. A good example of this is the Nottingham Prognostic Index (NPI)which is widely used in clinical practice [2] and takes the form: *NPI score = 0.2*Tumour size (cm) + Node Stage (1...3) + Histological Grade (1...3)*.

The same principles apply when flexible models are used, such as generic non-linear algorithms including artificial neural networks.

Paulo J.G. Lisboa

School of Computing and Mathematical Sciences, Liverpool John Moores University, Byrom Street, Liverpool L3 3AF, UK e-mail: P.J.Lisboa@ljmu.ac.uk

In the case of discrete time models of longitudinal data, the main variable is the event rate, also called the hazard rate

$$h(x_p, t_k) = P(t \leq t_k \mid t > t_{k-1}, x_i) \tag{4.1}$$

which is the probability that patient p with characteristics x_p survives to the end of time interval t_k given that the patient is known to have entered that interval without experiencing the event of interest. This is the output of the Partial Logistic Artificial Neural Network (PLANN) model [3]. This index is usually averaged over time to define a time-independent risk score over a predefined time interval.

The equivalent index to the NPI score is now the log-odds ratio

$$PLANN \; Risk \; Index = log\left(\frac{h(x_p, t_k)}{(1 - h(x_p, t_k))}\right) \tag{4.2}$$

where the conditional probability of class membership is directly estimated by the neural network output.

Once the risk score is defined, the population of patients at risk needs to be stratified for the purpose of tailoring adjuvant therapy and to enable comparisons between to be made between patient cohorts from different clinical centres, or subject to different clinical interventions, to be made between patients at similar risk by outcome. In survival analysis the most widely used statistic to test significant differences is the log-rank statistic.

The next section reviews current practice in the application of the log-rank test to stratification of patient data. This is followed by a case study of prognostic modelling of data from patients with operable breast cancer, comparing a statistical methodology (Cox regression) with PLANN-ARD [4] that uses the ARD framework to regularise the PLANN model and performed well in comparison with alternatives in a recent double blind benchmark of linear and generic non-linear survival models [5]. In section 4 the prognostic scores obtained on the same data by these two methods are stratified, comparing the standard application of the log-rank test with a novel methodology proposed in this chapter, to resolve significant issues of robustness in the allocation of patients into risk groups. In section 5 the cut-off points found with the novel methodology proposed are applied to a new cohort from the same centre where the training data set was collect, in order to evaluate the performance of the new methodology.

4.2 Application of the log-rank test to stratification of patient data

In the literature the approach to splitting risk indices into risk groups is not always stated clearly, sometimes stating the cut-off points of the respective risk scores without a clear indication of how these were obtained [6,7]. Where the split of the indices is at all explained, expert knowledge has been a factor as in the case for the widely used NPI. This index is designed for ease of use and is derived by rounding a more cumbersome Cox regression calculation, the cut-off points being chosen to best match the risk groups from the original model which, in turn, was split on the basis of best match with known clinical groups.

In another approach the indices are split into equal sized groups as suggested by Harrell *et al.* [8]. This tutorial in biostatistics suggests using deciles as a starting choice and in a prognostic model for ovarian cancer Clark *et al.* [9]used quartiles to partition the risk score.

A suggestion for an automated method is to use successive top-down splits by maximising the log-rank test statistic [10].

4.3 Prognostic modelling of breast cancer patients

The reference data set for this case study consists of routinely acquired clinical records for patients recruited by Christie Hospital, Manchester, during 1983-89. The specific cohort of interest is patients with early, or operable breast cancer, defined using the standard TNM (Tumour, Nodes, Metastasis) staging system as tumour size less than 5 cm, node stage less than 2 and without clinical symptoms of metastatic spread. This defines a case series (n=917) for a longitudinal cohort study with 5-year follow-up. The date of recruitment is the date of surgery and the event of interest is cancer specific mortality.

Earlier studies identified the following six predictive variables: age at diagnosis, node stage, histological type (lobular, ductal or in situ), ratio of axillary nodes affected to axilar nodes removed, pathological size (i.e. tumour size in cm) and oestrogen receptor count. All of these variables are banded and binary coded as 1-from-N. For details of the attribute assignment see [4].

Two analytical models were fitted to the data, starting with a piecewise linear model Cox regression, also known as proportional hazards. This model factorises dependence on time and the covariates, modelling the hazard rate for patient with clinical characteristics x_p at time t_k as follows:

$$\frac{h(x_p, t_k)}{1 - h(x_p, t_k)} = \frac{h_0(t_k)}{1 - h_0(t_k)} exp\left(\sum_{i=1}^{N_i} \beta x_i\right) \tag{4.3}$$

where h_0 denotes the empirical hazard for a reference population with covariate attributes all equal to zero and x_i is the static covariate vector. This was taken to be the standard used by the software that implemented the piecewise linear model Cox regression (SAS), where the reference population is the one which has the last attribute for each covariate. It was found that the risk group allocation is not sensitive to the choice of reference population. In contrast, the PLANN model is semi-parametric and with the following model for the hazard.

$$\frac{h(x_p,t_k)}{1-h(x_p,t_k)} = exp\left(\sum_{h=1}^{N_h} w_h \cdot g\left(\sum_{i=1}^{N_i} w_{ih} \cdot x_{pi} + w \cdot t_k + b_h\right) + b\right) \quad (4.4)$$

where the indices i and h denote the input and hidden node layers and the non linear function $g(\cdot)$ is a sigmoid.

Both models optimise the same objective function, namely the log-likelihood summed over the observed status of the patient sampled over of 60 months, with an indicator variable that is 1 for death attributed to breast cancer and 0 if the patient is observed alive. Using target values τ_{pk} as indicator labels and t_l as the time index.

$$G = -\sum_{p=1}^{No.patients} \sum_{k=1}^{t_l} [\tau_{pk} log\left(h(x_p,t_k)\right) + (1-\tau_{pk}) log\left(1-h(x_p,t_k)\right)] \quad (4.5)$$

Data for patients who are lost to follow-up are said to be right-censored and the patients no longer counts as part of the set of patients at risk.

In addition, over-fitting is avoided by regularisation, in the case of Cox regression using Akaike's Information Criterion (AIC) and in the neural network model with Automatic Relevance Determination (ARD) [11].

The prognostic index is defined in both cases as the covariate dependent term in eq. (2) and they are compared for the two models in fig. 4.1.

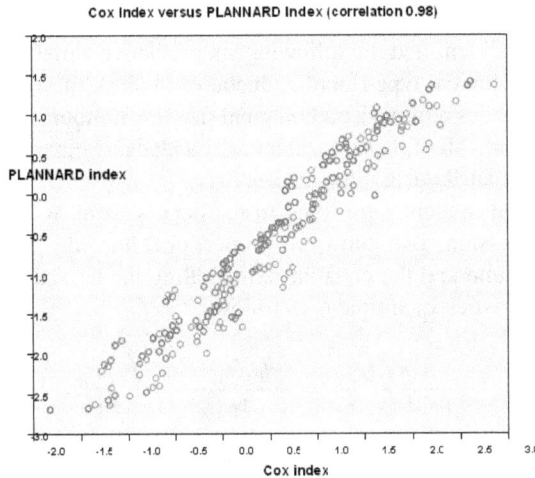

Fig. 4.1 Correlation between the prognostic derived with Cox regression and PLANN-ARD for a 5-year study of a patient cohort with early breast cancer. The proportionality of the hazards is borne out by the high correlation between the methods. In general, higher risk cohorts, longer follow-up times, studies of recurrence and models for other diseases will only be suitable for piecewise linear modelling if the proportionality of the hazards is observed.

The next stage in the modelling process is to use the prognostic index as the basis to partition the patient cohort into groups at similar risk of adverse outcome. This is the subject of the next section.

4.4 Robust methodology for stratification of severity of risk

The most widely used method for stratification of an empirical distribution of prognostic indices is to apply the log-rank test statistic from which the statistical significance for pairwise data partitions can be measured. Given that the test only applies in a pairwise manner, that is to say, for separating two cohorts at a time, this requires a search for the most appropriate threshold to divide the distribution of prognostic index scores.

An accepted strategy was implemented in SAS. It starts by sorting all the records by the value of the prognostic index. Next, the total data are divided into two groups at a threshold value that sweeps the full range of prognostic indices from minimum to maximum. For each threshold, the *log-rank statistic* is calculated and hence a *p-value* results. The maximum of the log-rank statistic determines the first cut-off point. The same method is then repeated in each of the separated cohorts until no further partitioning exceeds a pre-set confidence level which, for this study, is as p-value of 0.01 (99% of confidence), corresponding to a test statistic value of around seven.

In practice, the test statistic very much exceeds this value across a wide range of thresholds with the associated p-values forming a plateau indicating that there are a wide range of candidate cutpoints in addition to the maximum log rank statistic that has been selected as can be seen in fig. 4.2.

A new methodology is proposed to make the stratification of risk indices more robust. The new approach is bottom-up according to the following procedure which involves two nested loops:

Innor loop

i. Bin the risk indices into discrete intervals each containing a minimum number of cases (e.g. $n_{min}=10$).
ii. Calculate the log-rank statistic for each pair of adjacent cells and aggregate together the two cells with the smallest value of this test statistic.
iii. Repeat the process until the long-rank statistic is significant for all remaining cell pairs.

Outer loop

i. Draw a sample of the risk indices, with replacement, of size equal to the original data size – this is a bootstrap re-sample of the data.
ii. Apply the *inner loop* to convergence using the re-sampled data.
iii. Allocate each value in the full range of the risk index to a risk group, from $1..N_{groups}$

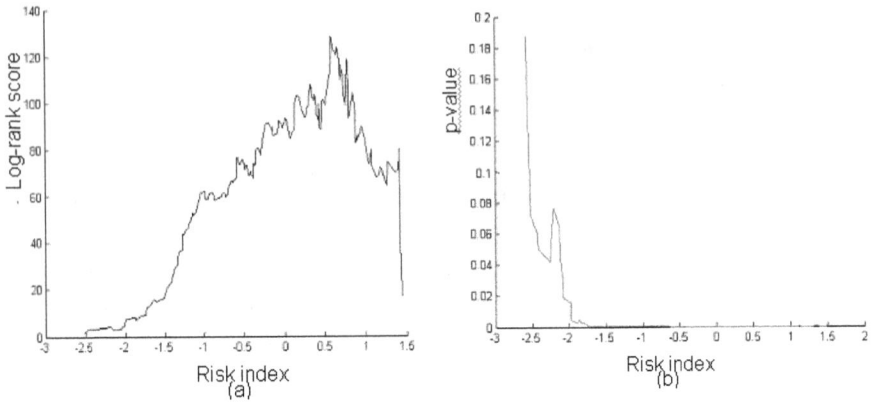

Fig. 4.2 The significance of data partitions in the top-down approach that is generally applied to stratify patient data in medical statistics detects the global maximum in (a), but this does not take into account that the statistical significance is high for a wider range of possible cut-off thresholds as shown in (b).

iv. Repeat from i a given number of times (e.g. $n_{resamples} = 3000$).
v. Identify the distribution of values of N_{groups} and discard all group assignments different from the mode of this distribution.
vi. For each value in the full range of the risk index, build a distribution of risk group allocations – this clearly indicates the cases that fit firmly into a risk group and those that are near the boundary between adjacent groups.
vii. Allocate each case in the original sample to the mode of the distribution of risk groups.

In the current case study, the risk group distributions obtained by this method are plotted in fig. 4.3.

The robustness of this approach to risk group identification is illustrated in fig. 4.4. The small size sample causing the unexpected outcome profiles in the solution with 6 risk groups may be an indication that this methodology is over-fitted to the training data.

4.5 Methodology out-of-sample Validation

After defining the risk score for each patient, based on the defined model, and stratifying the population of patients in different survival group risks, these need to be validated in order to evaluate their performance. There are several issues where prognostic models may not perform well and consequently risk stratification might fail, because there are deficiencies in standard modelling methods, such as the choice of variables to enter in the model; deficiencies in the design of prognostic studies as

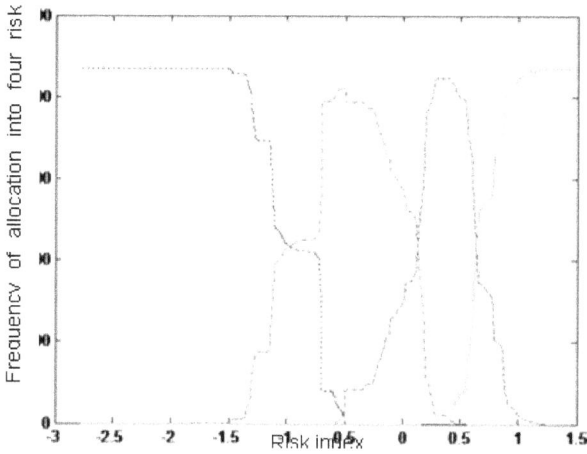

Fig. 4.3 Results from the proposed methodology for allocating risk groups from a risk index for severity of illness. Note that the group allocation frequencies vary smoothly, in contrast with the spot values of the log-rank test statistic in fig. 4.2.

what to do with missing data or inadequate sample size; or the models may not be transportable as the patients in different centres can have a very high dissimilarity. Given these reasons, it is very important to validate the proposed methodology with internal, temporal or external validation.

In our study, an out-of sample or temporal validation was done using a second data set from the same centre, which was collected between 1990 and 1993. Analysing the "Kaplan Meier" curves as well as log rank pairwise comparisons for the validation data set it can be evaluated the performance of the new methodology for stratification of illness indices.

	Risk Groups	1 x^2 (sig.)	2 x^2 (sig.)	3 x^2 (sig.)	4 x^2 (sig.)	5 x^2 (sig.)
Log Rank (Mantel-Cox)	2 x^2 (sig.)	0.0006 (0.9803)				
	3 x^2 (sig.)	6.3154 (0.0120)	3.8427 (0.0500)			
	4 x^2 (sig.)	12.2920 (0.0050)	4.3721 (0.0365)	0.1010 (0.7507)		
	5 x^2 (sig.)	32.7837 (<0.0001)	12.9783 (0.0003)	1.0443 (0.3068)	8.4239 (0.0037)	
	6 x^2 (sig.)	134.8670 (<0.0001)	50.1400 (<0.0001)	10.9272 (0.0009)	91.5980 (<0.0001)	15.3687 (<0.0001)

	Risk Groups	1 x^2 (sig.)	2 x^2 (sig.)	3 x^2 (sig.)
Log Rank (Mantel-Cox)	2 x^2 (sig.)	7.9833 (0.0047)		
	3 x^2 (sig.)	69.5909 (<0.0001)	36.7121 (<0.0001)	
	4 x^2 (sig.)	126.9977 (<0.0001)	73.5007 (<0.0001)	7.2322 (0.0072)

Table 4.1 Long rank pairwise comparisons for the validation data set. Both tables represent the validation modelling with PLANN-ARD. The top table represents the standard method and the bottom table represents the proposed method for increasing robustness in the risk stratification.

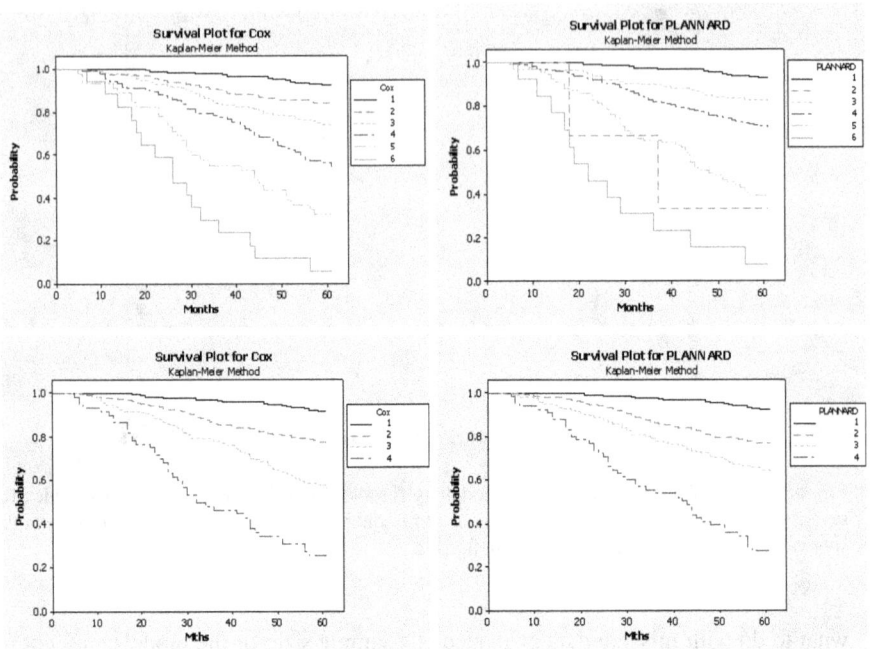

Fig. 4.4 Actuarial estimates of survival obtained with the Kaplan-Meier method, for the same cases (n=917), stratified using the log-rank test over a 60 month period. The top row uses the standard method and the bottom row uses the proposed method for increasing robustness in the risk stratification. The left column uses Cox regression modelling and the right column the PLANN-ARD neural network. The two modelling algorithms should be consistent, shown in fig. 1 but this is only apparent when the bootstrap method was applied.

The robustness of the new approach to risk group identification applied to an out-of-sample data set is illustrated in fig. 4.5 and table 4.1. Although the standard methodology applied to the training data set stratifies the patients with a significant difference in survival, the same it does not happen when it is applied to the validation data set.

The standard methodology's lack of robustness can be observed in the Kaplan Meier curves using the PLANN-ARD neuronal network for modelling (fig.4.5), where there is no significant survival difference between risk groups 1 and 2 and risk groups 3 and 4. The same evidence can be observed in the log-rank pairwise comparisons (Table 4.1), where there is no significant difference between the referred groups. On the other hand, both in Kaplan Meier curves and log-rank pairwise comparisons, it can be observed that there is a significant survival difference in patient stratification using the methodology proposed.

While analyzing the out-of-sample validation of both stratification methodologies, it can be concluded, as it was suggested before, that the standard method it is overfitted to the training data set, resulting in not so good patient stratification when

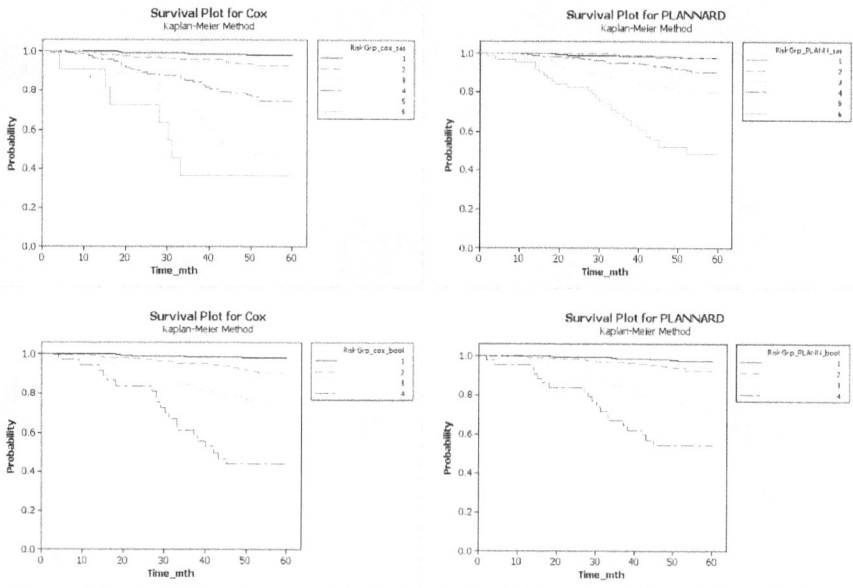

Fig. 4.5 Actuarial estimates of survival obtained with the Kaplan-Meier method, for the validation data set (931 cases), stratified using the cut-off points found in the training data set, over a 60 month period. The top row uses the standard method and the bottom row uses the proposed method for increasing robustness in risk stratification. The left column uses Cox regression modelling and the right column the PLANN-ARD neural network.

it is applied to a different data set. The new proposed methodology has however a very good performance when it is applied to an out-of-sample data set, as the difference in group-risk survival is highly significant.

4.6 Conclusions

The application of the log-rank test statistic to stratify patients by risk of adverse outcome is subject to variability due to the high prevalence of similar scores for many different risk thresholds. This results in unstable boundaries between strata, causing unwanted variability in allocation of patients into risk groups.

This chapter proposes a robust methodology for risk group allocation which exploits bootstrap re-sampling in order to stabilise the distribution of risk groups predicted for each value of the risk score index. The effectiveness and robustness of this methodology are shown by reference to a case study for operable breast cancer, using data from a longitudinal cohort study with 5-year follow-up.

In addition, the generic applicability of the proposed methodology is illustrated using both piecewise linear and neural network models of survival. While the results are consistent with earlier studies of the same data, the current findings are regarded as definitive on account of the robustness that has been added to the stratification process.

Moreover both methodologies were validated with an out-of-sample data set and the robustness of the new approach was confirmed, after evaluating the Kaplan-Meier curves and the log rank pairwise comparisons between the group risks.

Acknowledgments.

This work was carried out with partial funding from the Biopattern Network of Excellence FP6/2002/IST/1; N IST-2002-508803
(http://www.biopattern.org) and from the Fundação para a Ciência e Tecnologia, through the POS_C(SFRH/BD/30260/2006). Availability of data by Christie Hospital is gratefully acknowledged.

References

1. Cox, D. R. Regression models and life tables.: Journal of the Royal Statistical Society, B. 74,: 187-220 (1972)
2. Haybittle, J.L., Blamey, R.W., Elston, C.W., Johnson, J., Doyle, P.J., Campbell, F.C., Nicholson, R.I. and Griffiths, K.: A prognostic index in primary breast cancer. British Journal of Cancer. 45, 3621 (1982)
3. Biganzoli, E., Boracchi, P., Mariani, L. and Marubini, E.: Feed forward neural networks for the analysis of censored survival data: A partial logistic regression approach. Statistics in Medicine. 17, 1169-1186 (1998)
4. Lisboa, P.J.G., Wong, H., Harris, P. and Swindell, R.: A Bayesian neural network approach for modelling censored data with an application to prognosis after surgery for breast cancer. Artificial Intelligence in Medicine. 28, 1: 1-25 (2003)
5. Taktak, A., Antolini, L., Aung, M., Boracchi, P., Campbell, I., Damato, B., Ifeachor, E., Lama, N., Lisboa, P., Setzkorn, C., Stalbovskaya, V., and Biganzoli, E.: Double-blind evaluation and benchmarking of survival models in a multi-centre study. Comput. Biol. Med. 37, 8 (2007)
6. Guerra, I., Algorta, J., Diaz de Otazu, R., Pelayo, A., Farina, J.: Immunohistochemical prognostic index for breast cancer in young women. J Clin Pathol: Mol Pathol 56, 323–327 (2003)
7. Ortiz Sebastian S., Rodrıguez Gonzalez, J.M., Parilla Paricio, P., Sola Perez, J., Perez Flores, D., Pinero Madrona, A., Ramirez Romero, P., Tebar, F.J.: Papillary Thyroid Carcinoma: Prognostic Index for Survival Including the Histological Variety. Arch. Surg. 135 (2000)
8. Harrell, F.E., Lee, K.L., Mark, B.D.: Tutorial in Biostatistics Multivariate Prognostic Models: Issues in Developing Models, Evaluating Assumptions and Adequacy, and Measuring and Reducing Errors. Statistics in Medicine. 15, 361-387 (1996)
9. Clark. T.G., Stewart, M.E., Altman, D.G., Gabra, H., Smyth, J.F.: A prognostic model for ovarian cancer. Br J Cancer. 85, 944–52 (2001)
10. Williams, B.A., Mandrekar, J.N., Mandrekar, S.J., Cha, S.S. and Furth, A.F.: Finding Optimal Cutpoints for Continuous Covariates with Binary and Time-to-Event Outcomes. Technical Report Series #79, Mayo Clinic, Rochester, Minnesota, June 2006

11. MacKay, D.J.C.: Probable networks and plausible predictions – a review of practical Bayesian methods for supervised neural networks. Network: Computation in Neural Systems. 6: 469-505 (1995)

Chapter 5
Exploratory characterization of a multi-centre ^{1}H-MRS brain tumour database

Alfredo Vellido *, Margarida Julià-Sapé, Enrique Romero, and Carles Arús

Abstract Non-invasive techniques such as Magnetic Resonance Imaging (MRI) and Magnetic Resonance Spectroscopy (MRS) are often required for the diagnosis of tumours for which conclusive biopsies are not commonly available. While radiologists are used to interpreting MRI, many of them are not accustomed to make sense of the biochemical information provided by MRS. In this situation, oncology radiologists may benefit from the use of computer-based support in their decision making. As part of the AIDTumour research project, the analysis of MRS data corresponding to various tumour pathologies is used to assist expert diagnosis. The high dimensionality of the MR spectra might obscure atypical aspects of the data that would jeopardize their automated classification and, as a result, the process of computer-based diagnostic assistance. In this study, we put forward a method to overcome this potential problem that combines methods of visualization through non-linear dimensionality reduction, automatic outlier detection, and radiologists' expert opinion.

5.1 Introduction

Evidence-based medicine (EBM) has been defined as "healthcare practice that is based on integrating knowledge gained from the best available research evidence, clinical expertise, and patients' values and circumstances" [1], and its practice re-

Alfredo Vellido

Dept. de Llenguatges i Sistemes Informàtics - Universitat Politècnica de Catalunya, C. Jordi Girona, 1-3. 08034, Barcelona, Spain, http://www.lsi.upc.edu/~websoco/AIDTumour, e-mail: avellido@lsi.upc.edu, eromero@lsi.upc.edu

* A. Vellido is a Spanish Ministry of Science and Innovation (MINCINN) Ramón y Cajal program fellow researcher. The authors acknowledge funding from M.E.C. projects TIN2006-08114 and SAF2005-03650, and, from 1st January 2003, Generalitat de Catalunya (grant CIRIT SGR2005-00863).

quires health management to be based on objective findings, rather than on beliefs or subjective interpretation of the knowledge-base. The evidence available to the medical practitioner can take different and very heterogeneous forms. Here, we are specifically interested in quantitative information in the form of patients' biological signal as stored in an international, multi-centre, web-accessible database.

Decision making in oncology is a sensitive matter, and even more so in the specific area of brain tumour oncologic diagnosis, for which the direct and indirect costs - both human and financial - of misdiagnosis are very high. In this area, in which most diagnostic techniques must be non-invasive, clinicians should benefit from the use of an at least partially automated computer-based medical Decision Support System (DSS).

AIDTumour (Artificial Intelligence Decision Tools for Tumour diagnosis [10]) is a research project for the design and implementation of a medical DSS to assist experts in the diagnosis of human brain tumours on the basis of data obtained by Magnetic Resonance Spectroscopy (MRS). This is a technique that can shed light on cases that remain ambiguous after clinical investigation. The MRS data used in AIDTumour and analyzed in this paper belong to a complex multi-centre set containing cases of several brain tumour pathologies [11]. These data have undergone a rigorous pre-processing quality control that validates them from the viewpoint of the radiologists. Nevertheless, and for their use in an automated computer-based DSS, the various origins of these spectra and the complexity of their pre-processing make further data exploration advisable.

It might be problematic to include some of the spectra in an automated DSS without further ado for three different reasons: Firstly, some may contain measurement or acquisition artifacts that, even if not completely precluding diagnosis by visual inspection, might induce errors in computer-based diagnosis: these are what we call here *artifact-related outliers*. Secondly, atypical cases that do not contain artifacts but are nevertheless unrepresentative of the main distributions of the whole dataset: herein, these will be referred to as *distinct outliers* [4]. Thirdly, some cases with a clear biopsy-based diagnosis (tumour type attribution) may yield spectra that are quantitatively similar to those of other tumour types, misleading a computer-based classification system. Even if representative of the data as a whole, they are still unrepresentative of their own tumour type: these we will call *class outliers*. Note that these three reasons are not always mutually exclusive.

The aforementioned cases are likely to unduly bias the automated classification process in the DSS, even if for different reasons. The MRS dataset used in AIDTumour has already been used in the past for classification using simple linear techniques such as Linear Discriminant Analysis [5] or more sophisticated machine learning non-linear techniques such as Support Vector Machines [6, 7]. Nevertheless, it should be useful to address the potential problems outlined above as a preliminary step to classification. In this situation, EBM can be thought of as a framework to deal with medical uncertainty in a diagnostic setting, as its operational principles can be used to answer the questions for which there is good evidence and narrow the questions for which evidence is lacking [8]. This is the context in which Machine Learning (ML) and related methods can play a useful role [9].

In this study, we show the effectiveness of a method to identify and characterize potentially conflicting MRS data that combines techniques of non-linear dimensionality reduction, exploratory visualization, and outlier detection, with expert knowledge. The introduction of the latter is paramount, as it will help to skim those cases truly conflictive out of those shortlisted by blind quantitative criteria. Overall, this method is conceived as a preliminary step to data classification in the DSS. Dimensionality reduction is not trivial in this setting, as the available MRS data are scarce and high dimensional. Sammon's mapping [10] is used to this end. Generative Topographic Mapping (GTM [11]), a manifold learning model, is used to quantify spectra atypicality [4].

The remaining of the chapter is structured as follows. First, the MRS dataset is briefly described. This is followed, in section 3, by a description of the different analytical methods, in different subsections. Experimental results are presented in section 4, exhaustively describing problematic cases and illustrating some of them. The chapter closes with a section summarizing our conclusions.

5.2 MRS Data

The analysed MRS data correspond to 217 short-echo time (SET: PRESS 30-32 ms) and 195 long-echo time (LET: PRESS 135-144 ms) single voxel ^{1}H MR spectra acquired in vivo from brain tumour patients. They include 58 (SET) and 55 (LET) meningiomas (*mm*), 86 (SET) and 78 (LET) glioblastomas (*gl*), 38 (SET) and 31 (LET) metastases (*me*), 22 (SET) and 20 (LET) astrocytomas grade II (*a2*), 6 (SET and LET) oligoastrocytomas grade II (*oa*), and 7 (SET) and 5 (LET) oligo-dendrogliomas grade II (*od*).

These data are extracted from a web-accessible database [11] resulting from the *International Network for Pattern Recognition of Tumours Using Magnetic Resonance* (INTERPRET) European research project [12]. The criteria for the selection of cases to be included in the database were: a) that the case had a single voxel SET, 1.5 T spectrum acquired from a nodular region of the tumour; b) that the voxel was located in the same region as where subsequent biopsy was obtained; c) that the short-echo spectrum had not been discarded because of acquisition artifacts or other reasons; and d) that a histopathological diagnosis was agreed among a committee of neuropathologists. In those cases in which the spectra were obtained from normal volunteers without the pathology, or corresponded to abscesses or clinically proven metastases, biopsy was not required. For further details on data acquisition and processing, and on database characteristics, see, for instance, [13] and [11].

Class labelling was performed according to the World Health Organization (WHO) system for diagnosing brain tumours by histopathological analysis of a biopsy sample. For the reported analysis, spectra were bundled into three groups, namely: G1: *low grade gliomas* (*a2*, *oa* and *od*); G2: *high grade malignant tumours* (*me* and *gl*); and G3: *meningiomas* (*mm*). This type of grouping is justified [5] by the well-known difficulty in distinguishing between metastases and glioblastomas, due to their similar spectral pattern produced by the highly necrotic nature of these tumours. The clinically-relevant regions of the spectra were sampled to obtain 195 frequency intensity values (measured in parts per million (ppm), an adimensional unit of relative frequency position in the data vector), from 4.25 parts per million (ppm) down to 0.56 ppm, which become data attributes in the reported experiments.

5.3 Methods

5.3.1 MRS Data Visualization Through Sammon's Mapping

A typical *desiderata* for the visual representation of data and knowledge can be formulated in terms of maximizing structure preservation, as usual in multidimensional scaling [14]. In order to allow the visualization of the data through dimensionality reduction, the spectra were mapped onto a 3-D space through Sammon's mapping [10]. The non-linear mapping is constructed as to minimize the inter-point distortions it introduces, quantified by Sammon's error measure:

$$\frac{1}{\sum_{i<j}\delta_{ij}}\sum_{i<j}\frac{(\delta_{ij}-\xi_{ij})^2}{\delta_{ij}}, \tag{5.1}$$

where δ_{ij} is the Euclidean distance between spectra i and j in the original data space and ξ_{ij} is the Euclidean distance between the projections of these spectra in the 3-D space. A low Sammon error means that distances in the original space are preserved in the 3-D visualization space.

In this study, the minimization of the Sammon's error was performed by the Newton method. A collection of models was obtained by varying the initial points (100 different random values) and the step size (9 different values), for a total of 900 runs. The models with lowest Sammon's error were selected for further analysis.

5.3.2 Outlier Detection Using t-GTM

Generative Topographic Mapping (GTM [11]) is a non-linear latent variable model generating a mapping from a low dimensional latent space onto the multivariate data space. The mapping is carried through by a set of basis functions generating a constrained mixture density distribution. It is defined as a generalized linear regression model:

$$\mathbf{y} = \phi(\mathbf{u})\mathbf{W}, \tag{5.2}$$

where ϕ are M basis functions $\phi(\mathbf{u}) = (\phi_1(\mathbf{u}), ..., \phi_M(\mathbf{u}))$. For continuous data of dimension D, the choice in the original formulation were spherically symmetric Gaussians $\phi_m(\mathbf{u}) = \exp\{-1/2\sigma^2\|\mathbf{u}-\mu_m\|^2\}$, with centres μ_m and common width

σ; $\mathbf{W}$ is a matrix of adaptive weights w_{md} that defines the mapping, and $\mathbf{u}$ is a point in latent space. To avoid computational intractability, a regular grid of K points $\mathbf{u}_k$ can be sampled from the latent space. Each of them is mapped, using (5.2), into a low-dimensional manifold non-linearly embedded in the data space. Therefore, GTM can be considered as a manifold learning model. A probability distribution for the multivariate data $\mathbf{X} = \{\mathbf{x}_n\}_{n=1}^N$ can then be defined, leading to the following expression for a log-likelihood:

$$L = \sum_{n=1}^N \ln\left\{ \frac{1}{K} \sum_{k=1}^K \left(\frac{\beta}{2\pi}\right)^{D/2} \exp\left\{\frac{-\beta\|\mathbf{y}_k - \mathbf{x}_n\|^2}{2}\right\} \right\} \tag{5.3}$$

where a prototype $\mathbf{y}_k$ residing in the observed data space is obtained for each latent space point $\mathbf{u}_k$, using (5.2); and β is the inverse of the noise model variance. As for Finite Mixture Models, of which GTM is a manifold-constrained instance, the Expectation-Maximization (EM) algorithm is an straightforward alternative to obtain the Maximum Likelihood estimates of the adaptive parameters of the model, namely $\mathbf{W}$ and β.

For the standard Gaussian GTM, the presence of outliers is likely to negatively bias the estimation of the adaptive parameters. In order to overcome this limitation, the GTM was recently redefined [4] as a constrained mixture of Student's t distributions: the t-GTM, aiming to increase the robustness of the model towards outliers. The mapping described by Equation (5.2) remains, with the basis functions now being Student's t distributions and leading to the definition of the following mixture density:

$$p(\mathbf{x}|\mathbf{W},\beta,\nu_k) =$$
$$\frac{1}{K}\sum_{k=1}^K \frac{\Gamma(\frac{\nu_k+D}{2})\beta^{D/2}}{\Gamma(\frac{\nu_k}{2})(\nu_k\pi)^{D/2}}\left(1 + \frac{\beta}{\nu_k}\|\mathbf{y}_k - \mathbf{x}_n\|^2\right)^{\frac{\nu_k+D}{2}} \tag{5.4}$$

where $\Gamma(\cdot)$ is the gamma function and the parameter $\nu = (\nu_1,\ldots,\nu_K)$ represents the degrees of freedom for each component k of the mixture, so that it can be viewed as a tuner that adapts the level of robustness (divergence from normality) for each component.

As a byproduct of this reformulation of GTM, a statistic quantifying to what extent t-GTM considers a data case $\mathbf{x}_n$ to be an outlier can be defined, following [16], as $O_n = \sum_k p(\mathbf{u}_k|\mathbf{x}_n)\beta\|\mathbf{y}_k - \mathbf{x}_n\|^2$. The larger the value of this statistic, the more likely the case is to be an outlier. Notice that $p(\mathbf{u}_k|\mathbf{x}_n)$ is the responsibility assumed by a latent point k for the data case n and, the same as for the standard GTM, it is obtained as part of the Maximum Likelihood estimation of the model's parameters, in the M-step of the EM algorithm.

5.3.3 Shortlisting Outlier Cases of Interest

The process of shortlisting outlier cases of potential interest is structured in four stages:

- Sammon's mapping, as described in section 2, is first used to produce a nonlinear dimensionality reduction of the data to three dimensions.
- The free software package KING [15] is then used to visualize in 3-D the Sammon's mapping of the spectra, enabling a preliminary data exploration.
- The data projections obtained with Sammon's mapping are then modelled by t-GTM (using all the spectra for finding *artifact-related outliers* and *distinct outliers*, but only spectra belonging to, in turn, *G1*, *G2* and *G3* for finding *class outliers*), obtaining a value of O_n for each data case that quantifies the corresponding degree of atypicality. Histograms of O_n were generated to shortlist potentially conflictive cases of the three types described in the introduction. Loose thresholds of the statistic were set for the selection of the lists of outlier candidates.
- Using all this information, an expert in MRS then singled out those spectra she/he considered to be truly atypical in any sense and compared them to the characteristic spectra corresponding to their tumour type. When found to be *artifact-related outliers*, the spectrum was tagged in the database with information about their causes and recommendations on the suitability of its use for classification are made. When found to be a *distinct outlier*, it was tagged as such. When considered to be a *class outlier*, a warning was included in the tag so that it could be taken into account before attempting classification.

5.4 Experimental Results and Discussion

5.4.1 Short Echo Time ^{1}H MRS Data

The visualization of the high-dimensional spectra through Sammon's non-linear mapping is illustrated in Fig. 5.1. High grade malignant tumours are displayed in black, low grade gliomas in white, and meningiomas in gray. Overall, these three groups look well-defined and show a reasonable degree of separation, but it is also clear that some cases do not conform to this behaviour and that some of the issues outlined in the introduction should be considered.

Several SET cases of interest are also displayed in Fig. 5.2 for illustration: I1283 (a meningioma, which the expert described as being contaminated by noise, and affected by bad water suppression, polispiculated artifact and eddy currents), I0354 (a glioblastoma, which the expert described as being affected by a polispiculated artifact), I0135 (a glioblastoma that shows an extreme necrotic lipid pattern - a high peak at around 1.3 ppm - and a baseline artifact), and I0179 and I0450 (artifact-free

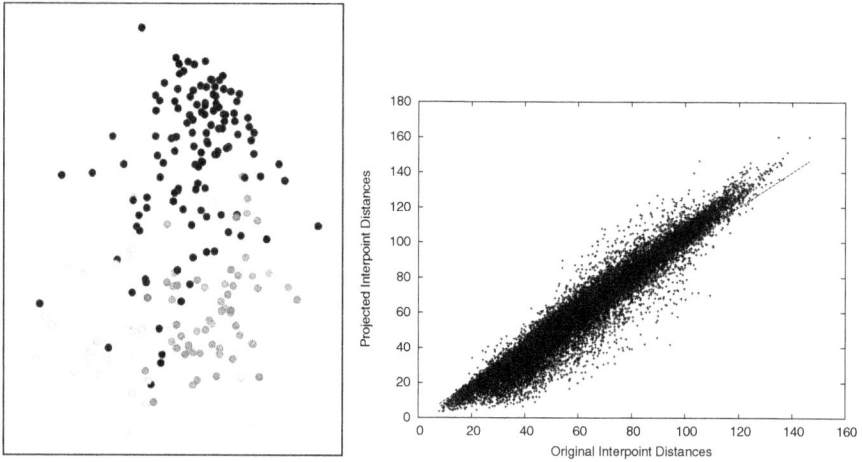

Fig. 5.1 A 3-D view of the projected MRS data obtained by Sammon's mapping (left: High grade malignant tumours are displayed in black, low grade gliomas in white, and meningiomas in gray) and the original and projected interpoint distances (right).

low grade gliomas with an unusual pattern of mobile lipids, with peaks at 0.8 and 1.3ppm as in *high grade malignant* tumours, but, in contrast to them, with the 0.8 ppm peak of the same magnitude as the 1.3 ppm peak).

The histogram in Fig. 5.3 displays the distribution of values of the statistic O_n, first calculated for the complete SET MRS dataset. A threshold of $O_n = 20$ was set to shortlist outlier candidate spectra. This yielded 23 potential outliers, which were inspected by an expert who decided that 19 of them (4 *distinct outliers* and 15 *artifact-related outliers*) qualified as such, for different causes listed in Table 1 (left). Notice that there are plenty of *low grade gliomas* (37% of all outliers, while only 16% of all data). Six different types of artifacts were found in the data, namely: spectra heavily contaminated by noise; bad water signal suppression as part of the data acquisition and/or pre-processing; incorrect spectrum alignment of the ppm reference; incorrect baseline; the existence of polispiculated artifact; and signal distorsion due to eddy currents (induced as a result of field gradient switching in signal acquisition).

As previously mentioned, spectra can also be atypical specifically with respect to their group of tumours. These are what we call *class outliers*. The histograms of O_n for each group of tumours are displayed in Fig.5.4. Six *low grade gliomas*, 20 *high grade malignant tumours*, and 13 *meningiomas* where shortlisted and inspected by the expert, who considered that, out of these, none of the *low grade gliomas*, only 9 *high grade malignant tumours*, and 8 *meningiomas* should be tagged as *class outliers*. Some of them also contain artifacts, given that, as mentioned in the introduction, *artifact-related outliers* and *class outliers* are not mutually exclusive characterizations. They are described in Table 2 (left). It is very interesting that, even

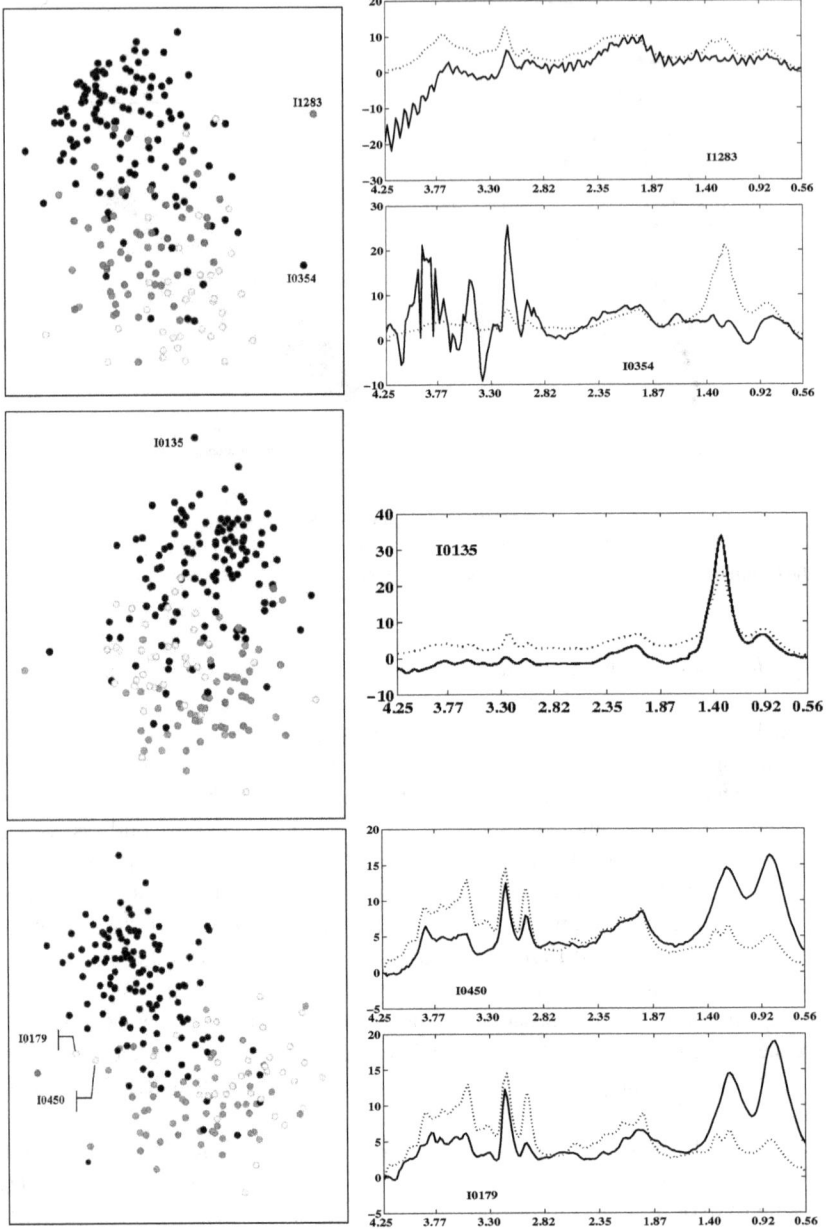

Fig. 5.2 3-D Sammon's mapping view of several cases of interest (with groups of tumours displayed in different shades of gray), on the left column, and their corresponding individual SET spectra (solid lines) and mean spectra (dotted lines) of the tumour groups they belong to, on the right column (case numbering as coded in the original INTERPRET database [12]). The abscissa axis displays frequency in ppm.

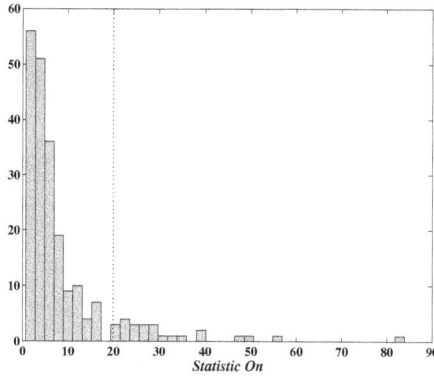

Fig. 5.3 Histogram of statistic O_n for the complete SET dataset. The selected threshold at value 20 is represented as a vertical dotted line.

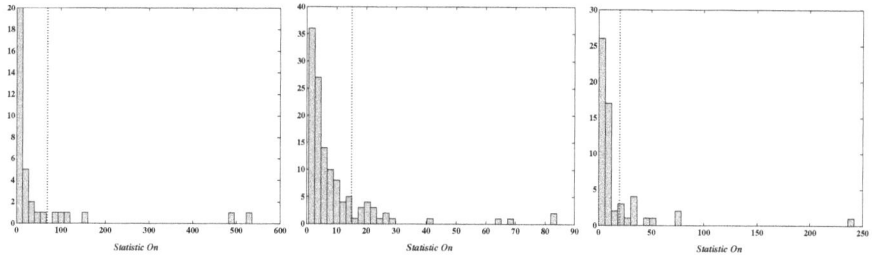

Fig. 5.4 Histogram of statistic O_n for each of the tumour groups in the SET dataset. (Left): *Low grade gliomas*; (centre): *High grade malignant tumours*; (right): *Meningiomas*. The selected thresholds are represented as vertical dotted lines.

though *low grade glioma* outliers are plentiful, as seen in Table 1 (left), there is no *class outlier* amongst them in the SET spectra, suggesting a well-defined against the rest but less-than-compact structure in this group of tumours.

5.4.2 Long Echo Time ^{1}H MRS Data

The histogram in Fig. 5.5 displays the distribution of O_n for the complete LET MRS dataset. A threshold of $O_n = 15$ was set to shortlist outlier candidate spectra. This yielded 21 potential outliers, which were again inspected by an expert, who decided that 18 of them qualified as such (3 *distinct outliers* and 15 *artifact-related outliers*). The corresponding characterization is presented in Table 1 (right). Interestingly, in this case there is almost no *low grade glioma* outlier and, instead, *high grade malignant* outliers predominate (67% of all outliers, while only 56% of all data).

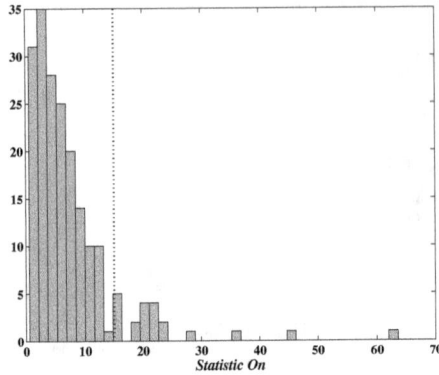

Fig. 5.5 Histogram of statistic O_n for the LET dataset. The selected threshold at value 15 is represented as a vertical dotted line.

Table 5.1 Outlier characterization of the SET (left) and LET (right) ^{1}H MRS datasets. Columnwise, *Id* is an anonymized case identifier from the INTERPRET database [12]; star superscripts indicate that there are artifacts that do not preclude the expert's correct interpretation of the case. *Tum* refers to tumour type (see labels in section 2). *Dis* refers to *Distinct outliers*. Six types of artifacts were found: *noi* stands for noise; *wat*, for bad water signal suppression; *ali*, for alignment; *lin*, linebase; *pol*, for the polispiculated artifact; and *edd* for eddy currents. See main text for details.

Id	Tum	Dis	Artifact-relat. outl.					
			noi	wat	ali	bas	pol	edd
I0335	G1(a2)	X						
I1052*	G1(a2)		X					
I1087*	G1(a2)			X				
I0060	G1(oa)		X					
I0069	G1(oa)					X		
I0450	G1(oa)	X						
I0179	G1(od)	X						
I0135*	G2(gl)					X		
I0172*	G2(gl)			X		X		
I0354*	G2(gl)						X	
I0421*	G2(gl)			X				
I1024*	G2(gl)			X				
I0055	G2(me)		X					
I0244*	G3(mm)	X						
I0375	G3(mm)		X					
I0381*	G3(mm)			X				
I0390*	G3(mm)						X	
I0393*	G3(mm)		X			X	X	
I1283*	G3(mm)		X	X			X	X

Id	Tum	Dis	Artifact-relat. outl.					
			noi	wat	ali	bas	pol	edd
I1061	G1(a2)				X			
I0062*	G2(gl)			X	X	X		
I0105*	G2(gl)	X						
I0172	G2(gl)				X	X		
I0175*	G2(gl)			X			X	
I0354*	G2(gl)				X		X	
I0428*	G2(gl)				X		X	
I1044*	G2(gl)						X	
I1057*	G2(gl)			X	X		X	
I1379*	G2(gl)			X				X
I0027	G2(me)			X		X		
I0368*	G2(me)				X		X	
I1070	G2(me)	X						
I0390*	G3(mm)						X	
I0420	G3(mm)						X	
I1074	G3(mm)			X				
I1090	G3(mm)	X						
I1378	G3(mm)			X			X	

We now turn our attention to LET *class outliers*, whose corresponding histograms for statistic O_n are displayed in Fig.5.6. Nine *low grade gliomas*, 7 *high grade malignant tumours*, and 10 *meningiomas* were shortlisted and inspected by

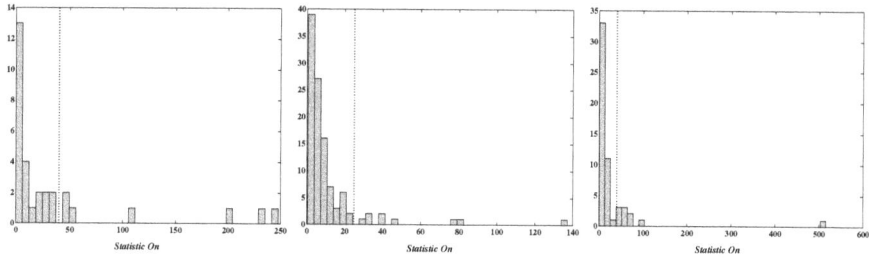

Fig. 5.6 Histogram of statistic O_n for each of the tumour groups in the LET dataset. (Left): *Low grade gliomas*; (centre): *High grade malignant tumours*; (right): *Meningiomas*. The selected thresholds are represented as vertical dotted lines.

the expert, who considered that, out of these, none of the *low grade gliomas*, only 2 *high grade malignant tumours*, and 5 *meningiomas* should be tagged as *class outliers*. Some of them also contain artifacts, and they are characterised in Table 2 (right). It is worth noting that there are far less *class outliers* in the LET dataset than in the SET one, suggesting a much more compact definition of the tumour groups in the former representation. It is also interesting that, again, there is no *class outlier* amongst the *low grade gliomas*. Together with the almost complete lack of outliers in this tumour group shown in Table 1 (right), this indicates that they have a much more compact and well-defined structure in the LET representation.

5.5 Conclusion

In this study, we have defined a general method for the identification and characterization of potentially conflicting MR spectra corresponding to an international, multi-center database of brain tumour pathologies. This method combines nonlinear dimensionality reduction, exploratory visualization, and automatic outlier detection techniques with expert knowledge. This combination of data-based analysis and human expertise is one of the distinctive hallmarks of Evidence-Based Medicine (EBM) for healthcare practice [1]. We project to embed this method in a medical DSS resulting from the AIDTumour [10] research project.

Some research questions remain unanswered and would require further research. The usefulness of outlier detection and characterization and its impact on automated tumour diagnostic classification should be assessed. For instance, the number of *class outliers* without artifacts is higher at SET than at LET. A straightforward interpretation could be that LET data are in general more homogeneous and therefore preferable for classification tasks. However, an expert spectroscopist may instead interpret that, with SET, more class outliers are detected (for example glioblastomas) due to the increased sensitivity of this echo time to mobile lipid presence. In any case, this brings to the fore the problem of subclass discovery: due to the inhomo-

Table 5.2 *Class outlier* characterization of the SET (left) and LET (right) ^{1}H MRS datasets, by groups of tumours. Label description as in Table 1.

Id	Tum	Artifacts					
		noi	wat	ali	bas	pol	edd
Low grade gliomas (G1)							
Ø							
High grade malignant (G2)							
I0021*	gl						
I0358*	gl					X	
I0200*	gl					X	
I1390	gl		X				
I1098*	gl						
I0168*	me						
I1076*	me				X		
I0352*	me				X		
I1377*	me						
Meningiomas (G3)							
I0160*	mm						
I1090*	mm						
I1073*	mm					X	
I0009	mm						
I0390*	mm					X	
I1378*	mm				X		
I0375	mm	X				X	
I1149*	mm						

Id	Tum	Artifacts					
		noi	wat	ali	bas	pol	edd
Low grade gliomas (G1)							
Ø							
High grade malignant (G2)							
I0105*	gl						
I1070	me						
Meningiomas (G3)							
I0114*	mm						
I1090	mm						
I1378	mm	X				X	
I0002*	mm						
I0009*	mm						

geneity of some tumour pathologic subtypes, it may make sense grouping distinct outliers, even from different pathologic subtypes, in separate, new subclasses.

Acknowledgments.

Authors gratefully acknowledge the former INTERPRET (EU-IST-1999-10310) European project partners. Data providers: Dr. C. Majós (IDI), Dr.À. Moreno-Torres (CDP), Dr. F.A. Howe and Prof. J. Griffiths (SGUL), Prof. A. Heerschap (RU), Dr. W. Gajewicz (MUL) and Dr. J. Calvar (FLENI); data curators: Dr. A.P. Candiota, Ms. T. Delgado, Ms. J. Martín, Mr. I. Olier and Mr. A. Pérez (all from GABRMN-UAB). C. Arús and M. Julià-Sapé are funded by the CIBER of Bioengineering, Biomaterials and Nanomedicine, an initiative of the *Instituto de Salud Carlos* III (ISCIII) of Spain.

References

1. Dickersin, K., Straus, S.E., Bero, L.A.: Evidence Based Medicine: Increasing, not Dictating, Choice. Brit. Med. J. 334(supl.1), s10 (2007)
2. Artificial Intelligence Decision Tools for Tumour diagnosis research project, http://www. lsi.upc.edu/~websoco/AIDTumour
3. Julià-Sapé, M., Acosta, D., Mier, M., Arús, C., Watson, D., The INTERPRET Consortium: A Multi-Centre, Web-Accessible and Quality Control-Checked Database of in Vivo MR Spectra of Brain Tumour Patients. Magn. Reson. Mater. Phy. MAGMA 19, 22–33 (2006)
4. Vellido, A., Lisboa, P.J.G.: Handling Outliers in Brain Tumour MRS Data Analysis through Robust Topographic Mapping. Comput. Biol. Med. 36, 1049–1063 (2006)
5. Tate, A.R., Majós, C., Moreno, A., Howe, F.A., Griffiths, J.R., Arús, C.: Automated Classification of Short Echo Time in Vivo ^{1}H Brain Tumor Spectra: a Multicenter Study. Magn. Reson. Med. 49, 29–36 (2003)
6. Devos, A., Lukas, L., Suykens, J.A.K., Vanhamme, L., Tate, A.R., Howe, F.A., Majós, C., Moreno-Torres, A., van der Graaf, M., Arús, C., Van Huffel, S.: Classification of Brain Tumours using Short Echo Time ^{1}H MR Spectra. J. Magn. Reson. 170, 164–175 (2004)
7. Luts, J., Heerschap, A., Suykens, J.A.K., Van Huffel, S.: A combined MRI and MRSI based multiclass system for brain tumour recognition using LS-SVMs with class probabilities and feature selection. Artif. Intell. Med. 40, 87–102 (2007)
8. Ghosh, A.K.: On the Challenges of using Evidence-Based Information: The Role of Clinical Uncertainty. J. Lab. Clin. Med. 144(2), 60–64 (2004)
9. Vellido, A., Biganzoli, E., Lisboa, P.J.G.: Machine Learning in Cancer Research: Implications for Personalised Medicine. In: 16th European Symposium on Artificial Neural Networks (ESANN 2008), In press. pp.55-64. d-Side pub., Evere, Belgium (2008)
10. Sammon Jr, J.W.: A nonlinear mapping for data structure analysis. IEEE T. Comput. C-18, 401–409 (1969)
11. Bishop, C.M., Svensén, M., Williams, C.K.I.: The Generative Topographic Mapping. Neural Comput. 10(1), 215–234 (1998)
12. INTERPRET project: http://azizu.uab.es/INTERPRET
13. INTERPRET project, Data Protocols: http://azizu.uab.es/INTERPRET/cdap.html
14. T. F. Cox and M. A. A. Cox, *Multidimensional Scaling*. Chapman and Hall, 2001.
15. KING visualization software, http://kinemage.biochem.duke.edu/software/ king.php
16. Peel, D., McLachlan, G.J.: Robust mixture modelling using the t distribution, Stat. Comput. 10, 339–348, (2000)

Chapter 6
Feature and Model Selection in ^{1}H-MRS single voxel spectra for cancer classification

Félix F. González-Navarro and Lluís A. Belanche-Muñoz

Abstract Machine learning is a powerful paradigm within which to analyze ^{1}H-MRS spectral data for the classification of tumour pathologies. An important characteristic of this task is the high dimensionality of the involved data sets. In this work we apply specific feature selection methods in order to reduce the complexity of the problem on two types of ^{1}H-MRS spectral data: long-echo and short-echo time, which present considerable differences in the spectrum for the same cases. The experimental findings show that the feature selection methods enhance the classification performance of the models induced by several off-the-shelf classifiers and are able to offer very attractive solutions both in terms of prediction accuracy and number of involved spectral frequencies.

6.1 Introduction

In vivo nuclear proton magnetic resonance spectroscopy (^{1}H-MRS) is a powerful technique that helps to observe metabolic processes in living tissue [2]. Although these metabolic functions are not fully understood, it is possible to employ *machine learning* techniques for the diagnosis and grading of adult brain tumours [3]. Several recent examples in the literature use machine learning techniques for distinguishing between different brain tumour pathologies (e.g. [4], [5]). Due to the high dimensionality (near 200 spectral measurements in the present study), these efforts use dimensionality reduction methods (feature selection and/or extraction) to lower the complexity of the problem. Of added practical importance is the interpretability of the solutions in terms of the obtained spectral frequencies, which limits the appli-

Lluís A. Belanche-Muñoz

Dept. de Llenguatges i Sistemes Informàtics, Universitat Politècnica de Catalunya, Ω-Building, North Campus. Barcelona, Spain. e-mail: belanche@lsi.upc.edu

cability of methods such as Principal Component Analysis (PCA) or Independent Component Analysis (ICA).

In a previous work, the ^{1}H-MRS long-echo data were analysed by the authors with the purpose of obtaining classification models showing good generalization ability after a strong dimensionality reduction process [1]. In particular, we used an Entropic Filtering Algorithm (EFA) for feature selection as a fast method to generate a relevant subset of spectral frequencies. Performance of other fast feature selection algorithms was also reported. In the present study we are interested in performing a more in-depth feature selection study in both long- and short-echo types of data by the introduction of bootstrap resampling techniques to yield mean performance estimates and their variability. It is important to point out that long-echo and short-echo time spectral points are considered as two separate sets of features. The first goal is to obtain simple models (in terms of low numbers of features) that generalize well. The second goal is to progress in the direction of assessing the differential performance for the two types of spectra, which present notable differences for the same cases.

6.2 An entropic filtering algorithm

Mutual Information (MI) measures the mutual dependence of two random variables. It has been used with success as a criterion for feature selection in machine learning tasks. In this work we use this concept embedded in a fast algorithm that computes MI between a set of variables and the class variable by generating first a "super-feature", obtained considering the concatenation of each combination of possible values of its forming features. In symbols, let $X = \{X_1, ..., X_n\}$ be the original feature set and consider a subset $\tau = \{\tau_1, \cdots, \tau_k\}$. A single feature $\mathscr{V}_\tau$ can be obtained uniquely, whose possible values are the concatenations of all possible values of the features in τ. The conditional entropy between $\mathscr{V}\tau$ and the class feature Y is then:

$$H(Y|\tau_1, \cdots, \tau_k) = H(Y|\mathscr{V}_\tau) = -\sum_{v \in \mathscr{V}_\tau} \sum_{y \in Y} p(v,y) log \frac{p(v,y)}{p(y)}. \qquad (6.1)$$

Proceeding in this way, the MI can be determined as a simple bivariate case: $I(\mathscr{V}_\tau; Y) = H(Y) - H(Y|\mathscr{V}_\tau)$. An *index of relevance* of the feature $X_i \in X$ to a class Y with respect to a subset $\tau \subset X$, inspired on [6], is given by:

$$R(X_i; Y|\tau) = \frac{I(X_i; Y|\mathscr{V}_\tau)}{H(Y|\mathscr{V}_\tau)} = \frac{H(Y|\mathscr{V}_\tau) - H(Y|X_i; \mathscr{V}_\tau)}{H(Y|\mathscr{V}_\tau)}. \qquad (6.2)$$

This measure $R(X_i; Y|\tau)$ can be regarded as a conditioned *coefficient of constraint* [7]. It takes values between zero (no relevance) and one (maximum relevance). Note that $H(Y|\mathscr{V}_\tau)$ and $H(Y|X_i; \mathscr{V}_\tau)$ can be small when the size of τ increases. This way of calculating feature subset relevance is used to evaluate subsets of spectra, embedded into a fast filter forward-search strategy, conforming the *Entropic Filtering Algorithm* (EFA), detailed next.

Let $D_{p\times(n+1)} = (d_{i,j})$ be a discrete data matrix described by n variables $X = \{X_1,\ldots,X_n\}$ (plus the class variable Y, in column $n+1$). The matrix D is first sorted using lexicographical order, which accelerates future computations (this is done only once). The code in Algorithm 1 incrementally computes conditional entropy of the class variable given a super-feature and can be implemented in only one pass over the observations (the symbol $\uplus$ stands for concatenation). Note that a practical implementation is straightforward because the matrix is ordered and therefore the involved rows are consecutive and always start from the current position.

Algorithm 1: Conditional Multivariate Entropy

Function H $(Y,\ \tau' \subseteq X)$

$v^- \leftarrow \uplus\{d_{1,j} \mid X_j \in \tau'\}$; $y^- \leftarrow d_{1,n+1}$
$cv \leftarrow 1; cy \leftarrow 1$
$H \leftarrow 0$
for $i \leftarrow 2$ **to** p **do**
$\quad v \leftarrow \uplus\{d_{i,j} \mid X_j \in \tau'\}$; $y \leftarrow d_{i,n+1}$
$\quad$ **if** $v^- = v$ **and** $y^- = y$ **then**
$\quad\quad cv \leftarrow cv + 1$
$\quad\quad cy \leftarrow cy + 1$
$\quad$ **else if** $v^- \neq v$ **then**
$\quad\quad H \leftarrow H + \frac{cy}{p}\log\frac{cy}{cv}$
$\quad\quad cv \leftarrow 1$
$\quad\quad cy \leftarrow 1$
$\quad$ **else**
$\quad\quad t \leftarrow |\{v \mid v = d_{l,j}, l = i,\ldots,p, X_j \in \tau'\}|$
$\quad\quad H \leftarrow H + \frac{cy}{p}\log\frac{cy}{t}$
$\quad\quad cv \leftarrow cv + 1$
$\quad\quad cy \leftarrow 1$
$\quad v^- \leftarrow v$
$\quad y^- \leftarrow y$
returns $-\left(H + \frac{cy}{p}\log\frac{cy}{cv}\right)$

Algorithm 2: Index of relevance R

Function R $(X_i \in X \setminus \tau,\ Y,\ \tau \subseteq X)$

returns $\dfrac{\mathbf{H}(Y,\tau) - \mathbf{H}(Y,\tau \cup \{X_i\})}{\mathbf{H}(Y,\tau)}$

Algorithm 3: Entropic Filtering Algorithm

$\Phi \leftarrow \emptyset$ /* Best Spectral Subset BSS */
repeat
$\quad \left| \begin{array}{l} x' \leftarrow \underset{x \notin \Phi}{\mathrm{argmax}}\{\mathbf{R}(x, Y, \Phi)\} \\ \Phi \leftarrow \Phi \cup \{x'\} \end{array} \right.$
until $R(Y, \Phi) = 1$ **or** $\Phi = X$;

In order to apply the algorithm, a discretization process is needed. Many dimensionality reduction studies use discretization schemes as a way to favor classification tasks (such as [8], [9]). This change of representation does not often result in a significant loss of accuracy (sometimes significantly improves it); it also offers large reductions in learning time. The CAIM algorithm [10] is selected because it is able to work with supervised data and does not require the user to define a specific number of intervals for each feature.

6.3 Experimental work

The echo time is an influential parameter in ^{1}H-MRS spectra acquisition. In short-echo time spectra (typically 20-40 ms) some metabolites are better evaluated (e.g. lipids, myo-inositol, glutamine and glutamate). However, there may be numerous overlapping resonances (e.g. glutamate/glutamine at 2.2 ppm and NAA at 2.01 ppm) which make the spectra difficult to interpret [11]. A long-echo time (270-288 ms) yields less metabolites but also less baseline distortion, resulting in a more readable spectrum. There are a few studies comparing the classification potential of the two types of spectra (see e.g. [11], [12]). These works seem to give a slight advantage to using short-echo time information or else suggest a combination of both types of spectra.
The analyzed ^{1}H-MRS dataset is detailed as follows:

- 266 single voxel *long-echo time* spectra acquired *in vivo* from brain tumour patients, out of which $p = 195$ are used in this study, including: meningiomas (55 cases), glioblastomas (78), metastases (31), astrocytomas Grade II (20), oligoastrocytomas Grade II (6) and oligodendrogliomas Grade II (5);
- 304 single voxel *short-echo time* spectra, of which $p = 217$ are used: meningiomas (58 cases), glioblastomas (86), metastases (38), astrocytomas Grade II (22), oligoastrocytomas Grade II (6) and oligodendrogliomas Grade II (7).

Class labelling was performed according to the World Health Organization system for diagnosing brain tumours by histopathological analysis of a biopsy sample. Both spectra were grouped into three super-classes: high-grade malignant tumours (metastases and glioblastomas), low-grade gliomas (astrocytomas, oligodendrogliomas and oligoastrocytomas) and meningiomas. The spectra consist of $n = 195$ frequency intensity values, from 4.21 ppm down to 0.51 ppm.

6.3.1 Experimental setup

Cross-validation (CV) has been used in estimating prediction errors in many statistical models such as classification, providing almost unbiased estimation. However, estimating misclassification error with small samples raises concerns over its performance since CV presents large variability.

Recent proposals include combining the bootstrap with CV (by performing CV on each of the bootstrap samples), thus conforming bootstrap CV or BCV [13]. Besides, bootstrap methods are well-suited for the construction of standard error estimates and confidence intervals when sample size is small or the distribution of the statistic is unknown. The ^{1}H-MRS data sets S were used to generate $B = 1000$ bootstrap samples $S_1,\dots,S_B$ that will play the role of *training sets*. Denote $T_i = S \setminus S_i$ the corresponding *test sets*. The training sets are used for the feature selection process itself (by the EFA), *a posteriori* classifier induction and model selection (by CV). The test sets are used to ascertain the generalization ability of the developed classifiers.

Six different classifiers were first designed using the training set by means of leave-one-out CV (LOOCV) and the full set of frequencies. The classifiers are: the nearest-neighbour technique with Euclidean metric (*kNN*) and parameter k (number of neighbours), the *Naïve Bayes classifier* (NB), a *Linear Discriminant classifier* (LDC), a *Quadratic Discriminant classifier* (QDC), *Logistic Regression* (LR) and a *Support Vector Machine with linear kernel* (lSVM) and parameter C (regularization constant). The EFA is applied to the discretized ^{1}H-MRS data (the training parts only) to obtain what will be called Best Spectral Subsets (BSS). Note that the EFA does not need an inducer. The classifiers are then built in the training sets using the original continuous frequencies (both in the full set and in the obtained BSSs) and evaluated in the corresponding test sets.

There is an inherent difficulty in applying a feature selection algorithm on every boostrap sample: the process yields a different (though probably quite similar) solution for every sample. Bias-variance for feature selection is an active field nowadays, but still there is no consensus on how to derive a single solution. In the present setting, this situation is aggravated since the EFA is capable of delivering *more than one solution*. This is because, in the last step of Algorithm 3, there may be more than one possibility of reaching maximum relevance. In this study it was decided to track them all. Hence, the application of the EFA yields a collection of solutions $\Sigma_1,\dots,\Sigma_B$, which are sets of feature subsets.

First every Σ_i was collapsed into a single subset σ_i by computing the following function:

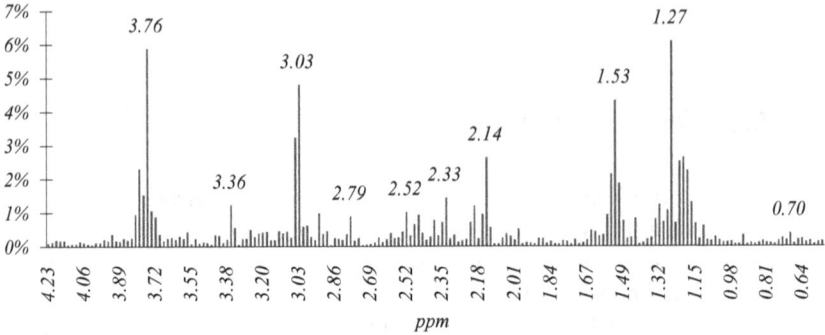

Fig. 6.1 Relative frequency distribution of spectral points selected by the EFA in the bootstrap samples in ^{1}H-MRS **Long** echo time data. Values on top of the peaks are set only as a reference.

$$\mathscr{I}(\Sigma) = \sum_{a \neq b \in \Sigma} MI(a,b) \tag{6.3}$$

where MI is the mutual information between features a and b and setting $\sigma_i = \operatorname{argmin}_{\Sigma \in \Sigma_i} \mathscr{I}(\Sigma)$. By construction of the EFA solutions, all the Σ_i are of the same size, so normalization is not necessary, given that all the summations in (6.3) have the same number of terms. This way of proceeding ensures that the chosen feature subset has maximum relevance (because it was one of the sets delivered by the EFA) and minimum redundancy among its features –because it is the minimizer of (6.3).

Once the boostrap feature sets $\sigma_1, \ldots, \sigma_B$ are obtained, we explored in this work four different strategies to obtain a single bootstrap solution σ^*, as follows. First create the set $\mathscr{F}$ as the union of all the σ_i. Define the *frequency* of a feature as the number of times it belongs to any of the σ_i, divided by B.

R=1   : Elements in $\mathscr{F}$ are fed into a forward selection algorithm sorted by descending frequency, where the stopping condition is reaching the maximum relevance ($R = 1$).

20% cum. : For many events, 80% of the effects (viz. classification ability) come from 20% of the causes (viz. spectral frequencies). Following this Pareto principle, spectral points in $\mathscr{F}$ were included in the final subset until reaching approximately 20% of the normalized cumulative frequency.

20% fea. : Similarly to the previous strategy, the 20% of the most frequent spectral points in $\mathscr{F}$ are included.

Peaks   : The elements in the sets $\sigma_1, \ldots, \sigma_B$ are considered as a distribution than can be plot. The most dominant peaks of the resulting histogram can be visually selected (see Figures 6.1 and 6.4).

In the next sections, for every feature selection experiment, the size of the corresponding BSSs, their test set performance, basic sample statistics and bootstrap

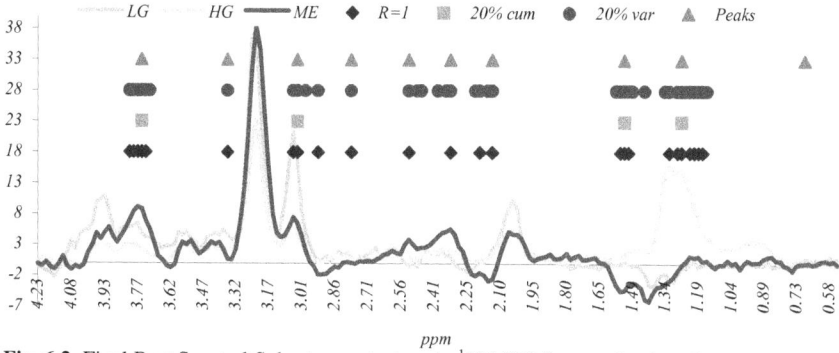

Fig. 6.2 Final Best Spectral Subsets per strategy in ¹H-MRS **Long** echo time data set.

confidence intervals (CI) are reported. This is done separately for the long-echo and short echo ¹H-MRS data sets.

6.3.2 Long-echo time experimental results

The positions in the spectrum of final long-echo time BSSs derived from the four strategies are reported in Table 6.1 and depicted in Figure 6.2, shown against average spectra per class. Five regions of spectral points can be observed: between 3.89 and 3.72 ppm Glutamate/glutamine-containing compounds and Alanine; another group is located at 3.03 ppm, having as center the Creatine; from 2.52 to 2.18 ppm the Glutamate and Glutamine metabolites can be identified; from 1.67 to 1.32 ppm the Alanine peak is roughly observed; and from 1.38 to 1.15 ppm Lactate is clearly identified.

Reduction	BSS	ppm
R=1	24	3.81, 3.79, 3.77, 3.76, 3.74, 3.36, 3.05, 3.03, 2.94, 2.79, 2.52, 2.33 2.20, 2.14, 1.55, 1.53, 1.51, 1.32, 1.29, 1.27, 1.23, 1.21, 1.19, 1.17
20% cum	4	3.76, 3.03, 1.53, 1.27
20% var	39	3.81, 3.79, 3.77, 3.76, 3.74, 3.72, 3.36, 3.05, 3.03, 3.00, 2.94, 2.79 2.52, 2.48, 2.46, 2.39, 2.35, 2.33, 2.22, 2.20, 2.16, 2.14, 1.57, 1.55 1.53, 1.51, 1.49, 1.44, 1.34, 1.32, 1.30, 1.29, 1.27, 1.25, 1.23, 1.21 1.19, 1.17, 1.15
Peaks	10	3.76, 3.36, 3.03, 2.79, 2.52, 2.33, 2.14, 1.53, 1.27, 0.70

Table 6.1 Final subsets of spectral points (features) obtained by each strategy on the ¹H-MRS **Long** echo time data set.

Complete test-set performances on the ^{1}H-MRS long-echo data set are displayed as follows. In Table 6.2 the five rows indicate the size of the used BSS and mean classification performance in the form $\mu^* \pm \frac{\sigma^*}{\sqrt{B}}$, where μ^* and σ^* are the test set mean and standard deviation for accuracy in the bootstrap samples, respectively. These figures give a first impression of mean test-set performance and its stability. Each row corresponds to a different method of choosing the BSS. Specifically, NR stands for 'no reduction', and the other four are the strategies described in previous sections.

Test set confidence intervals (shown in Table 6.3) can be obtained by the bootstrap percentile method, as follows: let $e^* = (e_1^*, \dots, e_B^*)$ denote the error obtained by a given configuration (classifier plus reduction strategy) on the bootstrap samples. The CI is constructed by ordering e^* in ascending order and choosing critical value observations as the endpoints of the confidence intervals. For instance, for $B = 1000$, observations 26 and 975 are the endpoints of the 95% CI.

Reduction	BSS	NB	kNN	LDC	QDC	LR	lSVM
NR	195	0.83±0.001	0.78±0.001	*	*	0.72±0.002	0.85±0.001
R=1	24	0.85±0.001	0.84±0.001	0.83±0.001	*	0.77±0.002	0.80±0.001
20% cum	4	0.82±0.001	0.80±0.001	0.82±0.001	0.83±0.001	0.84±0.001	0.84±0.001
20% fea	39	0.85±0.001	0.84±0.001	0.82±0.001	*	0.77±0.001	0.82±0.001
Peaks	10	0.82±0.001	0.84±0.001	0.83±0.001	0.79±0.001	0.83±0.002	0.84±0.001

Table 6.2 Bootstrap mean classification performance on the ^{1}H-MRS **Long** echo time test sets. Results marked with ($*$) indicate numerical problems (number of observations less or equal than number of features). NR stands for 'no reduction'.

Reduction	NB	kNN	LDC	QDC	LR	lSVM
NR	(0.75,0.90)	(0.69,0.87)	*	*	(0.62,0.81)	(0.76,0.92)
R=1	(0.77,0.92)	(0.76,0.91)	(0.75,0.90)	(0.37,0.78)	(0.66,0.85)	(0.71,0.88)
20% cum	(0.74,0.90)	(0.72,0.88)	(0.74,0.90)	(0.75,0.90)	(0.77,0.92)	(0.77,0.91)
20% fea	(0.77,0.93)	(0.75,0.91)	(0.74,0.90)	*	(0.67,0.85)	(0.74,0.90)
Peaks	(0.76,0.92)	(0.76,0.91)	(0.77,0.91)	(0.71,0.89)	(0.76,0.91)	(0.75,0.92)

Table 6.3 Confidence intervals by percentile method of bootstrap mean classification performance on the ^{1}H-MRS **Long** echo time test sets. Results marked with ($*$) indicate numerical problems (number of observations less or equal than number of features). NR stands for 'no reduction'.

Previous published work analyzing similar ^{1}H-MRS data used PCA followed by LDC to distinguish between high-grade malignant tumours and meningiomas, obtaining a mean AUC (area under the ROC curve) of 0.94, using 6 principal components [5]. The same method was used to distinguish between high-grade malignant tumours and astrocytomas Grade II (part of the low-grade gliomas super-class), obtaining a mean AUC of 0.92, also using 6 principal components. There is an interesting link with the present work in the low numbers of final dimensions. However, two drawbacks of PCA are that all the spectra participate in the linear combination, and the fact that the linear combination may mix both positive and negative weights, which might partly cancel each other. In [3], LDC with 6 spectral frequencies (3.72,

3.04, 2.31, 2.14, 1.51 and 1.20 ppm) achieved a 83% of correct classification on an independent test set, this time using exactly the same three super-classes that we have analyzed in this study. The reliability of this result is confirmed by looking at the obtained performance for LDC in Table 6.2. Other classifiers obtain slightly better generalization results with different subsets of spectra.

6.3.3 Short-echo time experimental results

The positions in the spectrum of final short-echo time BSSs derived from the four strategies are reported in Table 6.4 and depicted in Figure 6.3, shown against ave-rage spectra per class. It is observed that almost all the dominant metabolite peaks and their neighborhood are covered by the final subsets: Glutamate-Glutamine and Alanine in 3.79-3.72ppm region; Creatine at 3.03ppm; Glutamate-Glutamine and N-acetylaspartate between 2.46 and 2.05ppm; Alanine, Lactate and Lipids in 1.57 and 1.11ppm region.

Reduction	BSS	ppm
R=1	36	3.81, 3.79, 3.77, 3.64, 3.62, 3.60, 3.58, 3.55, 3.32, 3.26, 3.19, 3.17
		3.13, 3.05, 3.03, 2.96, 2.84, 2.43, 2.41, 2.37, 2.35, 2.31, 2.29, 2.27
		2.25, 2.14, 2.12, 1.38, 1.36, 1.34, 1.32, 1.30, 1.29, 1.27, 0.98, 0.87
20% cum	4	3.05, 3.03, 2.37, 1.27
20% var	39	3.81, 3.79, 3.77, 3.64, 3.62, 3.60, 3.58, 3.55, 3.32, 3.26, 3.19, 3.17
		3.05, 3.03, 2.96, 2.84, 2.81, 2.43, 2.41, 2.39, 2.37, 2.35, 2.33, 2.31
		2.29, 2.27, 2.25, 2.14, 2.12, 1.38, 1.36, 1.34, 1.32, 1.30, 1.29, 1.27
		1.02, 0.98, 0.87
Peaks	8	3.81, 3.60, 3.26, 3.03, 2.37, 2.12, 1.32, 0.98

Table 6.4 Final subsets of spectral points by each strategy on the ^{1}H-MRS **Short** echo time data set.

Complete test-set performances on the ^{1}H-MRS short-echo data set are displayed in a way analogous to those for the long-echo data set, for ease of comparison, in Table 6.5. Test set confidence intervals are shown in Table 6.6.

Reduction	BSS	NB	kNN	LDC	QDC	LR	ISVM
NR	195	0.82±0.001	0.81±0.001	*	*	0.68±0.002	0.85±0.001
R=1	36	0.85±0.001	0.87±0.001	0.86±0.001		0.82±0.001	0.85±0.001
20% cum	4	0.82±0.001	0.83±0.001	0.88±0.001	0.86±0.001	0.87±0.001	0.87±0.001
20% fea	39	0.85±0.001	0.87±0.001	0.86±0.001	*	0.80±0.001	0.84±0.001
Peaks	8	0.83±0.001	0.85±0.001	0.89±0.001	0.87±0.001	0.87±0.002	0.87±0.001

Table 6.5 Bootstrap mean classification performance on the ^{1}H-MRS **Short** echo time test sets. Results marked with (*) indicate numerical problems (number of observations less than number of features). NR stands for 'no reduction'.

Previous existing work analyzing the same ^{1}H-MRS data using 10 principal com-ponents and LDC as classifier obtained at most 85% of correct classification [4].

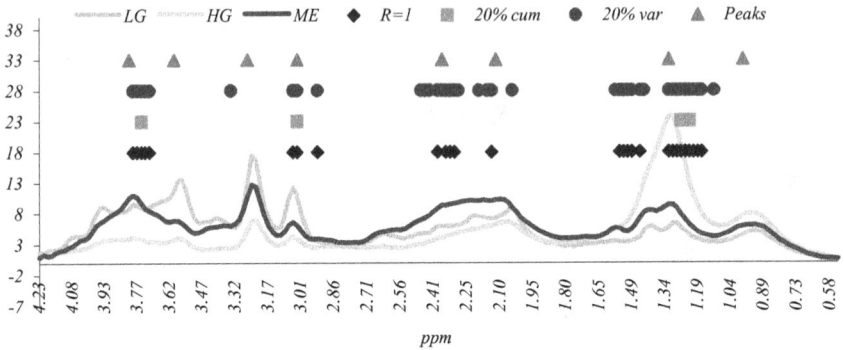

Fig. 6.3 Final Best Spectral Subsets per strategy in ^{1}H-MRS **Short** echo time data set.

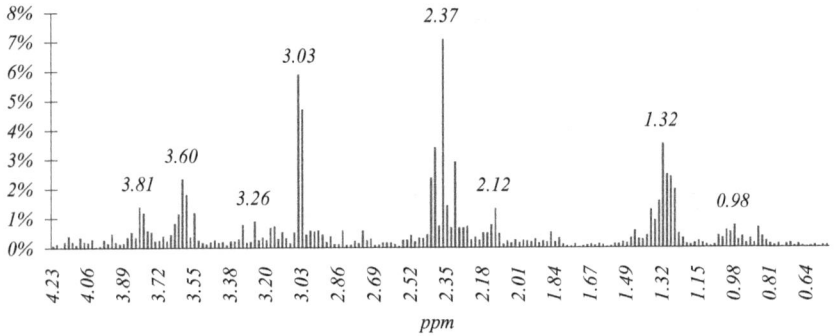

Fig. 6.4 Relative frequency distribution of spectral points selected by the EFA in the bootstrap samples in ^{1}H-MRS **Short** echo time data. Values on top of the peaks are set only as a reference.

Reduction	NB	kNN	LDC	QDC	LR	lSVM
NR	(0.75,0.89)	(0.73,0.88)	*	*	(0.59,0.78)	(0.77,0.91)
R=1	(0.78,0.91)	(0.80,0.93)	(0.78,0.92)	*	(0.73,0.89)	(0.78,0.92)
20% cum	(0.74,0.90)	(0.76,0.90)	(0.81,0.94)	(0.79,0.92)	(0.80,0.93)	(0.80,0.93)
20% fea	(0.78,0.92)	(0.80,0.93)	(0.79,0.92)	*	(0.72,0.88)	(0.77,0.91)
Peaks	(0.77,0.90)	(0.78,0.91)	(0.83,0.95)	(0.79,0.94)	(0.79,0.94)	(0.79,0.93)

Table 6.6 Confidence intervals by percentile method of bootstrap mean classification performance on the ^{1}H-MRS **Short** echo time test sets. Results marked with (∗) indicate numerical problems (number of observations less than number of features). NR stands for 'no reduction'.

Another study reported 89% accuracy (again using LDC as classifier) with 5 spectral frequencies (3.76, 3.57, 3.02, 2.35, 1.28 ppm) in an independent test set [3]. Similarly than for long-echo time data, the reliability of this result can be contrasted by looking at the obtained performances for LDC in Table 6.5 and specially with the CIs developed in Table 6.6.

6.3.4 Summary of findings

In view of these results, several remarks are in order, concerning experimental findings:

1. Feature selection appears as a viable avenue of dimensionality reduction: performance is very similar or slightly better to the 'no reduction' case for all the classifiers and reduction methods. This is specially so in the case of the *kNN* classifier, arguably the classifier that suffers the most from very high dimensionality. All the reduction methods, with a maximum of a fifth of spectral frequencies, obtain mean performance figures quite close or superior to using the full set. This behaviour is important in general, both for computational and scientific reasons.
2. All the tested classifiers seem to reach a consensus on performance that varies little with the reduction method and points to data characteristics as main limitation, given the very different nature of the classifiers. On the negative side, this prevents any of the classifiers being highlighted as the preferred one. However, given the necessary model selection step involved in the lSVM and the kNN, we recommend using LDC or NB, although this latter makes stronger data assumptions.
3. Among the different reduction methods, performance seems to be dependent on the chosen classifier, but in a precise way. The **R=1** and **20% fea** methods are more convenient for NB and kNN, while the **20% cum** and **Peaks** methods are more suited for LDC, LR and the lSVM, regardless of time information. Given the nature of the data, and the BSS sizes in Tables 6.2 and 6.5, it seems plausible to interpret that the former classifiers need more features to achieve a similar level of performance.
4. As stated previously, some existing works indicate an advantage in using short-echo time information [11], [12]. The present work adds support to this finding, since all the classifiers (with the single exception of NB, which shows anyway very similar figures) obtain slightly better results for short-echo time data.
5. The resulting sets of spectral frequencies (shown in Tables 6.1 and 6.4) may be subject of a more in-depth interpretation in terms of known metabolites. Of special importance is the clarity of relative frequency distribution of spectral points selected by the EFA in the bootstrap samples, for both echo times (Figs 6.1 and 6.4). This task falls beyond the scope of this paper.

6.4 Conclusions and future work

The problem of performing feature selection in two high-dimensional ^{1}H-MRS data sets has been studied. These data sets correspond to different brain tumours and are presented in both long- and short-echo time. An entropic filtering algorithm has been applied in a set of bootstrap samples combined with several methods for the obtention of a unique spectral subset. An added advantage of this method is the absence of any parameter tuning. The obtained subsets can then be used for developing classifiers using cross-validation in the bootstrap samples.

In all, the methods have provided a drastic dimensionality reduction while being competitive or sometimes improving on the performance of the full set of spectral frequencies, for both types of data. Comparative results with similar studies confirm previous results in terms of prediction accuracy, stressing the importance of careful and stable feature selection methods in this particular kind of data. Additional comments are made regarding the consistency of the different classifiers and the differential behaviour between long- and short-echo time. The introduction of resampling techniques as the bootstrap allows to yield more reliable performance estimates.

A specific and promising line of work will study the combination of both echo times of data acquisition, in the form of a single data set where the different spectral frequencies are put together. This poses a further computational challenge, given that the number of features will double while the number of observations will remain approximately the same. Some recent studies suggest that this combination of echo times may improve the classification results for the problem investigated. Notice that the most important spectral frequencies reported in Figs. 6.1 and 6.4 are quite different from each other (with the possible exception of the Creatine, which is common to both). It would therefore be sensible to argue that this process of data combination may permit the obtention of better subsets of frequencies.

A second possible avenue for further research would consist on the study of different tumour groupings, corresponding to either different pathologies to the ones analyzed here, or to different tumour type categorization criteria.

Acknowledgements

Authors gratefully acknowledge the former INTERPRET partners (INTERPRET, EU-IST-1999-10310) and, from 1st January 2003, Generalitat de Catalunya (grants CIRIT SGR2001-194, XT 2002-48 and XT 2004-51); data providers: Dr. C. Majós (IDI), Dr.A. Moreno-Torres (CDP), Dr. F.A. Howe and Prof. J. Griffiths (SGUL), Prof. A. Heerschap (RU), Dr. W. Gajewicz (MUL) and Dr. J. Calvar (FLENI); data curators: Dr. A.P. Candiota, Ms. T. Delgado, Ms. J. Martín, Mr. I. Olier, Mr. A. Pérez and Prof. Carles Arús (all from GABRMN-UAB). Authors also acknowledge funding for CICyT TIN2006-08114 and SAF2005-03650 projects; the Mexican CONACyT and Baja California University and thank the anonymous reviewers for their helpful suggestions.

References

1. González, F., Belanche, Ll.: Feature Selection in *in vivo* ^{1}H-MRS single voxel spectra. In Procs. of the KES 2008 conference, Zagreb, Croatia.
2. Sibtain, N.: The clinical value of proton magnetic resonance spectroscopy in adult brain tumours. Clinical Radiology **62** (2007) 109–119.
3. Tate, A.R., Underwood, J., Acosta, D.M., Julià-Sapé, M., Majós, C., Moreno-Torres, À., Howe, F.A., Van der Graaf, M., Lefournier, V., Murphy, M.M., Loosemore, A., Ladroue, C., Wesseling, P., Bosson, J.L., Cabåas, M.E., Simonetti, A.W., Gajewicz, W., Calvar, J., Capdevila, A., Wilkins, P.R., Bell, B.A., Rémy, C., Heerschap, A., Watson, D., Griffiths, J.R., Arús, C.: Development of a decision support system for diagnosis and grading of brain tumours using in vivo magnetic resonance single voxel spectra. NMR in Biomedicine **19** (2006) 411–434.
4. Ladroue, C.: Pattern Recognition Techniques for the Study of Magnetic Resonance Spectra of Brain Tumours. PhD thesis, St. George's Hospital Medical School, United Kingdom (2003).
5. Devos, A.: Quantification and classification of MRS data and applications to brain tumour recognition. PhD thesis, Katholieke University Leuven, Belgium (2005).
6. Wang, H.: Towards a unified framework of relevance. PhD thesis, University of Ulster, United Kingdom (1996).
7. Coombs, C. H., Dawes, R. M., and Tversky, A.: *Mathematical Psychology: An Elementary Introduction*, Prentice-Hall, Englewood Cliffs, NJ (1970).
8. Ng, M., Chan, L.: Informative gene discovery for cancer classification from microarray expression data. In: IEEE Workshop on Machine Learning for Signal Processing, IEEE (2005) 393–398.
9. Le, D., Satoh, S.: Robust object detection using fast feature selection from huge feature sets. In: 13th International Conference on Image Processing, IEEE (2006) 961–964.
10. Kurgan, L., Cios, K.: CAIM discretization algorithm. IEEE Transactions on Knowledge and Data Engineering **16**(2) (2004) 145–153.
11. Majós, C., Julià-Sapé, M., Alonso, J., Serrallonga, M., Aguilera, C., Acebes, J.J., Arús, C., Gili, J.: Brain tumor classification by proton MR spectroscopy: Comparison of diagnostic accuracy at short and long TE. American Journal of Neuroradiology **25** (2004) 1696–1704.
12. García-Gómez, J.M., Tortajada, S., Vidal, C., Julià-Sapé, M., Luts, J., Van Huffel, S., Arús, C., Robles, M.: On the use of long TE and short TE SV MR spectroscopy to improve the automatic brain tumor diagnosis. Technical report, In `ftp://ftp.esat.kuleuven.ac.be/pub/SISTA/ida/reports/07-55.pdf` (2007)
13. Fu, W., Carroll, R., Wang, S.: Estimating misclassification error with small samples via bootstrap cross-validation. Bioinformatics **21**(9):1979-1986 (2005).

Chapter 7
Rule-based assistance to brain tumour diagnosis using LR-FIR

Àngela Nebot *, Félix Castro, Alfredo Vellido, Margarida Julià-Sapé, and Carles Arús

Abstract This chapter describes a process of rule-extraction from a multi-centre brain tumour database consisting of nuclear magnetic resonance spectroscopic signals. The expert diagnosis of human brain tumours can benefit from computer-aided assistance, which has to be readily interpretable by clinicians. Interpretation can be achieved through rule extraction, which is here performed using the LR-FIR algorithm, a method based on fuzzy logic. The experimental results of the classification of three groups of tumours indicate in this study that just three spectral frequencies, out of the 195 from a range pre-selected by experts, are enough to represent, in a simple and intuitive manner, most of the knowledge required to discriminate these groups.

7.1 Introduction

Uncertainty is inherent to clinical oncology decision making, and poses a challenge for the development of any intelligent technologies with that purpose. The evidence, qualitative and quantitative, available to medical decision makers in oncology is growing exponentially. This situation justifies the design and development of computer-based decision support systems (DSS). The use of general medical DSS is now widespread and reasonably successful [1], but it is still rather uncommon to find any medical standard DSS using Computational Intelligence (CI) methods.

Àngela Nebot
Dept. de Llenguatges i Sistemes Informàtics - Universitat Politècnica de Catalunya, C. Jordi Girona, 1-3. 08034, Barcelona, Spain. http://www.lsi.upc.edu/ websoco/AIDTumour e-mail: angela@lsi.upc.edu, fcastro@lsi.upc.edu

* The authors acknowledge funding from Spanish M.E.C. projects TIN2006-08114 and SAF2005-03650, as well as, from 1st January 2003, Generalitat de Catalunya (grant CIRIT SGR2005-00863). Alfredo Vellido is a M.E.C. Ramón y Cajal researcher. Félix Castro is a research fellow within the PROMEP program of the Mexican Secretary of Public Education.

One of the potential drawbacks affecting the application of CI methods in general to the analysis of cancer data is the often limited interpretability of the results they yield. This is an extremely sensitive issue in a critical context such as oncology diagnosis. As stated in [2], "a DSS for medical diagnosis should support a comprehensible reasoning schema that corresponds to the human reasoning process". One way to overcome interpretability limitations, even though not the only one, is by explaining the operation of CI models using rule extraction methods. The interpretability of the model results should be greatly improved by their description in terms of reasonably simple and actionable rules that doctors and clinicians could rely on. In fact, rule extraction should provide clinicians, on whom the final responsibility for diagnosis rests, with an explanation about how a CI or related computer-based method has reached its decision [3].

Several authors have, in recent years, resorted to rule extraction from CI and related models in cancer research. Many of these involve the analysis of breast cancer data [4], although rule extraction for the classification of leukaemia and colon cancer data has also been proposed, for instance, in [5, 6]. Given that this paper is mostly concerned with fuzzy methods, it is worth noting that fuzzy theory for cancer analysis has been applied in conjunction with evolutionary algorithms in [7], Artificial Neural Networks in [5, 8] and rough sets in [9].

This paper describes a process of rule-extraction from a brain tumour database, and it is meant to be part of the design of the prototype medical DSS that is the goal of the AIDTumour (Artificial Intelligence Decision Tools for Tumour diagnosis [10]) research project. The multi-centre database under analysis in this study consists of Magnetic Resonance Spectroscopy (MRS) cases [11], corresponding to several tumour types. Rules to discriminate between tumours are extracted here using the Linguistic Rules in Fuzzy Inductive Reasoning (LR-FIR) algorithm [12]. At its core, FIR is a qualitative modeling and simulation methodology based on the observation of the input/output behavior of the system to be modeled, rather than on structural knowledge about its internal composition.

LR-FIR starts from the systems' model that the FIR method identifies in the form of pattern rules and, from them, is able to perform efficient generalization, deriving a set of actionable and realistic linguistic rules. The LR-FIR resulting rules preserve the system behaviour patterns and increase their interpretability.

The rest of the paper is organized as follows. The LR-FIR technique is described in section 2. The MRS dataset under study is briefly introduced in section 3. The rule extraction results are presented and discussed in section 4. The paper ends with a summary of conclusions and an outline of future research.

7.2 LR-FIR

The Fuzzy Inductive Reasoning (FIR) methodology is a mathematical tool for the modeling and simulation of complex systems. FIR is based on systems behavior rather that on structural knowledge. It is able to perform a selection of the system's

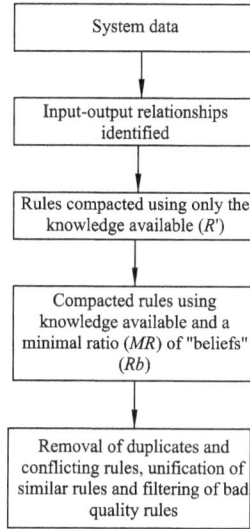

Fig. 7.1 Schematic representation of LR-FIR methodology

relevant variables and to obtain the causal and temporal relationships between them in order to infer the future behavior of that system. It also has the ability to describe systems that cannot easily be described by classical mathematics (e.g. differential equations), i.e. systems for which the underlying physical laws are not well understood. A FIR model is a qualitative, non-parametric, shallow model based on fuzzy logic, run under the Visual-FIR platform developed in Matlab®.

The FIR model consists of its structure (relevant variables) and a set of input/output relations (history behavior). Feature selection in FIR is based on the maximization of the models' forecasting power quantified by a Shannon entropy-based quality measure. Once the most relevant variables are identified, they are used to derive the set of input/output relations from the training data set. Detailed descriptions of the FIR methodology and Visual-FIR platform can be found in [13, 14]. The history behavior is subsequently used as input for the LR-FIR technique, which extracts and compacts the information contained on it.

The LR-FIR method aims to obtain interpretable, realistic and efficient rules, describing the behavior of the analyzed system. Predictive IF-THEN rules are chosen due to the fact that this kind of rules are very intuitive and are typically used for representing knowledge in the artificial intelligence area. Fig. 7.1 shows the main phases of LR-FIR.

The LR-FIR method can be summarized as a set of ordered steps:

1. *Basic compactation.* This is an iterative step that evaluates, one at a time, all the input/output relationships of a set R, which is compacted on the basis of the "knowledge" obtained by FIR. A subset R_c can be compacted in the form of a single rule r_c, when all premises P but one (P_a), as well as the consequence C, share the same values. Premises, in this context, represent the input features, whereas consequence is the output feature. If the subset contains all legal values LV_a of P_a, all these relationships can be replaced by a single rule, r_c, that has a value of -1 in the premise P_a. A -1 value means that this premise can take any of its possible values. When more than one -1 value, P_{ni}, is present in a compacted rule r_c, it is compulsory to evaluate the existence of conflicts by expanding all P_{ni} to all their legal values LV_a, and comparing the resultant rules X_r with the original relations R. If conflicts, Cf, exist, the compacted rule r_c is rejected, and otherwise accepted. In the latter case, the previous relationships subset, R_c is replaced by the compacted rule r_c. Conflicts occur when one or more extended rules, X_r have the same values in all its premises, P, but different values in the consequence C.
2. *Improved compactation.* Whereas the previous step only structures the available knowledge and represents it in a more compact form, the improved compactation step extends the knowledge base R to cases that have not previously been used to build the model. Thus, whereas step 1 leads to a compacted data base that only contains knowledge, the enhanced algorithm contains undisputed knowledge and uncontested belief, R_b. The improved compactation is an extension of the basic compactation, where a consistent and reasonable minimal ratio, MR, of the legal values LV_a should be present in the candidate subset R_c, to compact it in the form of a single rule r_c.

The obtained set of rules is subjected to a number of refinement steps: removal of duplicate rules and conflicting rules; unification of similar rules; evaluation of the obtained rules and removal of rules with low specificity and sensitivity values (this concepts are introduced later in the paper). For a more detailed description of LR-FIR methodology the user is referred to [12].

7.3 MRS datasets

The analyzed data correspond to 217 single-voxel, short echo-time ^{1}H-MR spectra acquired in vivo from brain tumour patients, classified according to the World Health Organization (WHO) system for diagnosing brain tumours by histopathological analysis of a biopsy sample. Three groups of tumours are analyzed in this study: G1, which includes 22 astrocytomas of grade 2, 6 oligoastrocytomas and 7 oligodendrogliomas, and is referred to as *low-grade gliomas*; G2, which includes 86 glioblastomas and 38 metastases, and is referred to as *high-grade malignant tumours*; and, finally, G3 includes 58 low-grade meningiomas. For each patient, there is a magnetic resonance spectrum consisting of 195 frequencies [11]. Each frequency (measured in parts per million (ppm), an adimensional unit of relative frequency position in the data vector) is treated as a data feature and, therefore, the dataset consists of 217 cases and 195 features. It was divided for analysis into training and test sets (balanced to account for class prevalence) containing, in turn, 163 and 54 patients.

7.4 Experimental results and discussion

FIR performs a selection of relevant variables in order to identify the structure of the model. In our experiments with the tumour groups described in the previous section, FIR found the frequencies 2.77ppm (herein referred to as f2.77), 2.33ppm (f2.33) and 1.34ppm (f1.34) to be the most relevant features. FIR also estimates the relative level of influence of the selected features on the discrimination of the tumour groups, which is 0.44 for f2.33, 0.43 for f1.34 and 0.13 for f2.77, (they add up to 1). In other words, f2.33 (corresponding to the presence of glutamate and macromolecules) and f1.34 (corresponding to lipids) are the most relevant frequencies in terms of the rules extracted for discriminating the tumour groups as described by the short echo-time MRS data. f2.77 (with no clear metabolic interpretation yet) also helps in the discrimination, but to a lesser extent and with a collateral role.

The three selected features are then used to compute the input/output relations, i.e. the history behavior that is the input information for the LR-FIR methodology. The rules obtained by LR-FIR for each of the three output classes, i.e. each group of brain tumours are presented in the first column of Table 1. A filtering threshold of 0.1 was applied, which means that the rules with specificity or sensitivity values lower or equal to 0.1 were deleted. Specificity is defined as one minus the ratio of the number of out-of-class data records that the rule identifies to the total number of out-of-class data. Sensitivity is the ratio of the number of in-class data that the rule identifies to the total number of in-class data. For simplicity, the antecedents of each rule are described in terms of numeric labels that correspond to intervals of values of the selected features, as described in Table 2. The second and third columns of Table 1 present the specificity and sensitivity metrics of the rules obtained for the training data set.

Table 7.1 *Brain tumours linguistic rules obtained for the training data set*

Rules	Spec.	Sens.
IF f2.33 IN 1-2 AND f1.34 IN 1-2 THEN CASE IN G1	0.72	0.77
IF f2.77 IN 2-3 AND f2.33 IN 3 AND f1.34 IN 1 THEN CASE IN G1	0.85	0.23
JOINT METRICS *Low grade gliomas*	**0.70**	**1**
IF f2.33 IN 1-2 AND f1.34 IN 2-3 THEN CASE IN G2	0.91	0.89
JOINT METRICS *High grade malignant*	**0.91**	**0.89**
IF f2.33 IN 3 AND f1.34 IN 1-2 THEN CASE IN G3	0.89	0.89
JOINT METRICS *Meningiomas*	**0.89**	**0.89**

Table 7.2 *Value intervals for each tumour group for variables: f2.77, f2.33 and f1.34*

	f2.77	f2.33	f1.34
Label 1	[-2.143...2.2336]	[0.16822...4.6769]	[-2.1358...9.3643]
Label 2	[2.2336...3.3045]	[4.6769...7.3388]	[9.3643...23.9711]
Label 3	[3.3045...10.4054]	[7.3388...16.1629]	[23.9711...37.1991]

The set of linguistic rules inferred (see Table 1) are then validated by computing the specificity and sensitivity metrics for the test data set. The evaluation results are presented in Table 3.

Table 7.3 *Specificity and sensitivity metrics obtained for the test data set*

Rules	Spec.	Sens.
IF f2.33 IN 1-2 AND f1.34 IN 1-2 THEN CASE IN G1	0.73	0.92
IF f2.77 IN 2-3 AND f2.33 IN 3 AND f1.34 IN 1 THEN CASE IN G1	0.85	0.077
JOINT METRICS *Low grade gliomas*	**0.71**	**1**
IF f2.33 IN 1-2 AND f1.34 IN 2-3 THEN CASE IN G2	0.92	0.93
JOINT METRICS *High grade malignant*	**0.92**	**0.93**
IF f2.33 IN 3 AND f1.34 IN 1-2 THEN CASE IN G3	0.98	0.92
JOINT METRICS *Meningiomas*	**0.98**	**0.92**

The task of rule extraction is to discover a set of rules from training data. Ideally, the obtained rules as a whole should explain not just the training instances but also unseen instances in the system. Our test results, reported in Table 3, comply with this requirement, as the specificity and sensitivity are still considerably high, and even better than the training ones. This confirms that the obtained set of rules represents the underlying behaviour of the system in a reliable way.

The obtained rules were analyzed and validated by MRS radiology experts. They agreed that the obtained results were intuitive, realistic, and mostly consistent with their own perception of the three cancer groups involved in the study.

For illustration, the meaning of the first rule for G1 in Table 1 and Table 2 is as follows: if, for a given case, the value (L2-normalized height) of feature f2.33 is in the intervals with label 1 or 2, and the value of feature f1.34 is in the intervals with labels 1 or 2, then the case is inferred to be a *Low grade glioma* (G1).

Our goal from the onset was to obtain simple and interpretable models that represent as accurately as possible the behaviour of the system. The resulting rules described in Table 1 and table 3 strike that balance: Only three frequencies of the spectra are enough to classify with an acceptable accuracy the three groups of tumours. Moreover, only two rules are needed to represent G1 quite accurately, while groups G2 and G3 only require a single rule for each one of them. This parsimonious representation of the three tumour groups under analysis should be extremely easy to act upon by medical experts. The interpretability of these rules can be further improved by representing them graphically on top of the characteristic spectra of the three types of tumours, as in Figs. 7.2, 7.3, and 7.4.

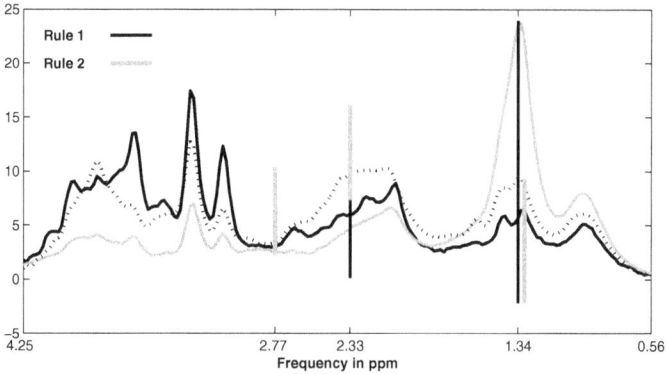

Fig. 7.2 Graphical representation, as vertical bars, of the rules described in Table 1 for *Low grade gliomas*, using the value intervals described in Table 2. They are described on top of the mean spectra for *Low grade gliomas* (G1, *solid black line*), *High grade malignant* (G2, *solid gray line*), and *Meningiomas*, (G3, *dotted black line*). *Vertical axis*: L2-normalized spectral intensity; *horizontal axis*: frequency chemical shift (ppm) with respect to water at 4.7 ppm.

The visual interpretation of the rules for *high-grade malignant tumours* (G2) and *meningiomas* (G3) is straightforward. The case of *low-grade gliomas* (G1) is more striking, specially the value intervals of f2.77 and f2.33 for rule 2, which might seem to correspond to G3 instead of G1. In fact, this rule is only describing two outlier spectra of oligoastrocytomas (hence its low sensitivity). Therefore, this should be considered as a spurious rule. It illustrates two different things at once: the usefulness of rule visualization and the negative effect of data outliers on automated classification.

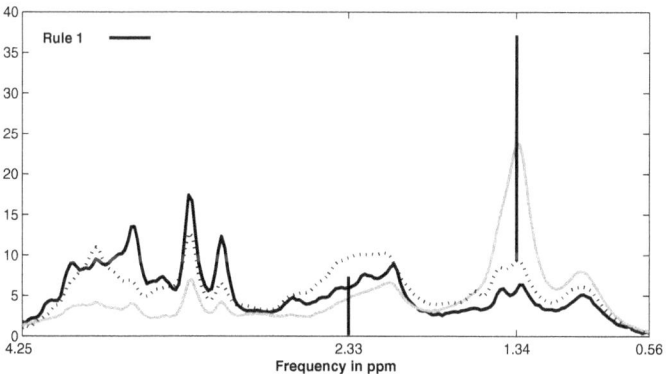

Fig. 7.3 Graphical representation, as in Fig. 7.2, of the rules described in Table 1 for *High grade malignant* tumours.

7.5 Conclusions

The interpretability of results is key in clinical oncology diagnostic assistance through computer-based methods. It has been shown in this paper, in a problem concerning the discrimination of diverse brain tumours using MRS data, that the interpretability of the problem can be greatly improved and simplified by its description in terms of a parsimonious set of simple and actionable rules that doctors and clinicians could rely on. The novel LR-FIR methodology has been used to this end in this study. LR-FIR is an hybrid fuzzy method that starting from the predictive model obtained by FIR is able to perform a knowledge generalization to derive a set of IF-THEN rules.

Three groups of tumours were analyzed in this study, namely low-grade gliomas, high-grade malignant tumours and meningiomas. For each of the 217 patients in the database, a MRS spectrum comprising 195 frequencies was available. The applied fuzzy methodology identified the frequencies f1.34, f2.33 and f2.77 as the most relevant for the classification of the brain tumours. This is a most interesting result, due to the very small size of the selected subset and because two of the frequencies have a direct metabolic correspondence that eases medical interpretation.

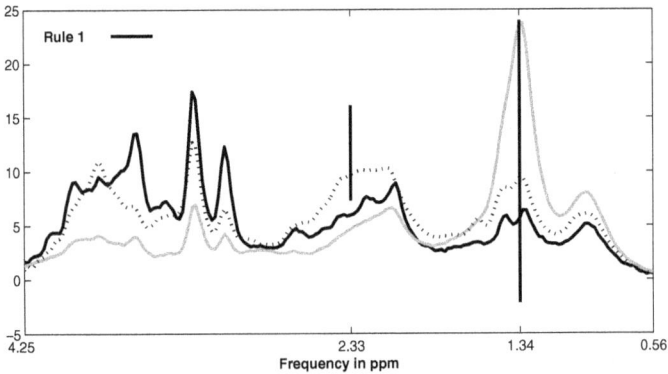

Fig. 7.4 Graphical representation, as in Fig. 7.2, of the rules described in Table 1 for *Meningiomas* tumour class.

The experimental results have shown that the rule extraction method presented in this paper was able to obtain comprehensible, actionable and realistic linguistic rules describing patient brain tumour pathologies. It has been shown that as few as two rules are enough to classify the low-grade gliomas with reasonable accuracy, while only one rule is needed for high-grade malignant tumours and meningiomas. The knowledge extracted from this set of rules could be used to improve tumour diagnostic assistance in a CI-based medical DSS tool.

Future research will focus on the integration of this rule extraction method and its results in the prototype medical DSS resulting from the AIDTumour [10] research project.

Acknowledgments.

Authors gratefully acknowledge the former INTERPRET (EU-IST-1999-10310) European project partners. Data providers: Dr. C. Majós (IDI), Dr.À. Moreno-Torres (CDP), Dr. F.A. Howe and Prof. J. Griffiths (SGUL), Prof. A. Heerschap (RU), Dr. W. Gajewicz (MUL) and Dr. J. Calvar (FLENI); data curators: Dr. A.P. Candiota, Ms. T. Delgado, Ms. J. Martín, Mr. I. Olier and Mr. A. Pérez (all from GABRMN-UAB). The CIBER of Bioengineering, Biomaterials and Nanomedicine is an initiative of the "Instituto de Salud Carlos III" (ISCIII) of Spain.

References

1. Garg, A.X., Adhikari, N.K.J., McDonald, H., Rosas-Arellano, M.P., Devereaux, P.J., Beyene, J., Sam, J., Haynes, R.B.: Effects of Computerized Clinical Decision Support Systems on Practitioner Performance and Patient Outcomes: A Systematic Review. J. Amer. Med. Assoc. 293, 1223–1238 (2005)
2. Tung, W.L., Quek, C.: GenSo-FDSS: a Neural-Fuzzy Decision Support System for Pediatric ALL Cancer Subtype Identification Using Gene Expression Data. Artif. Intell. Med. 33, 61–88 (2005)
3. Mitra, S.: Computational Intelligence in Bioinformatics. In: Peters, J.F., Skowron, A., van Albada, D. (eds.) Transactions on Rough Sets III. LNCS, vol. 3400, pp. 134–152. Springer, Heidelberg (2005)
4. Vellido, A., Lisboa, P.J.G.: Neural Networks and Other Machine Learning Methods in Cancer Research. In (eds.) IWANN 2007. LNCS, Vol. 4507, pp. 964–971. Springer, Heidelberg (2007)
5. Futschik, M.E., Reeve, A., Kasabov, N.: Evolving Connectionist Systems for Knowledge Discovery from Gene Expression Data of Cancer Tissue. Artif. Intell. Med. 28, 165–189 (2003)
6. Chen, Z., Lia, J., Wei, L.: A Multiple Kernel Support Vector Machine Scheme for Feature Selection and Rule Extraction from Gene Expression Data of Cancer Tissue. Artif. Intell. Med. 41, 161–175 (2007)
7. Peña-Reyes, C.A., Sipper, M.: A Fuzzy-Genetic Approach to Breast Cancer Diagnosis. Artif. Intell. Med. 17, 131–155 (1999)
8. Takahashi, H., Masuda, K., Ando, T., Kobayashi, T., Honda, H.: Prognostic Predictor with Multiple Fuzzy Neural Models Using Expression Profiles from DNA Microarray for Metastases of Breast Cancer. J. Biosci. Bioeng. 98, 193–199 (2004)
9. Hassanien, A.E.: Fuzzy Rough Sets Hybrid Scheme for Breast Cancer Detection. Image Vision Comput. 25, 172–183 (2007)
10. Artificial Intelligence Decision Tools for Tumour diagnosis (AIDTumour) research project, http://www.lsi.upc.edu/~websoco/AIDTumour
11. Julià-Sapé, M., et al.: A Multi-Centre, Web-Accessible and Quality Control-Checked Database of in Vivo MR Spectra of Brain Tumour Patients. Magn. Reson. Mater. Phy. MAGMA 19, 22–33 (2006)
12. Castro, F., Nebot, A.: Un Algoritmo para la Extracción Automática de Reglas Lógicas a partir de Modelos FIR. Technical Report, Universitat Politècnica de Catalunya LSI-07-7-R (2007).

13. Nebot, A., Cellier, F.E., Vallverdú, M.: Mixed Quantitative:Qualitative Modeling and Simulation of the Cardiovascular System. Comput. Meth. Prog. Bio. 55, 127-155 (1998)
14. Escobet, A., Nebot, A., Cellier,F.E. :Visual-FIR: A Tool for Model Identification and Prediction of Dynamical Complex Systems. Simul. Model. Pract. Th. 16, 76–92 (2008)
15. Vellido, A., Biganzoli, E., Lisboa, P.J.G.: Machine Learning in Cancer Research: Implications for Personalised Medicine. In: 16th European Symposium on Artificial Neural Networks (ESANN 2008), In press. d-Side pub., Evere, Belgium (2008)

Chapter 8
Statistical assessment of MSigDB gene sets in colon cancer

Angela Distaso, Luca Abatangelo, Rosalia Maglietta, Teresa Maria Creanza, Ada Piepoli, Massimo Carella, Annarita D'Addabbo, Sayan Mukherjee, and Nicola Ancona *

Abstract Gene expression profiling offers a great opportunity for understanding the key role of genes in alterations which drive a normal cell to a cancer state. A deep understanding of the mechanisms of tumorigenesis can be reached focusing on deregulation of gene sets or pathways. To this end, we measure the amount of deregulation and assess the statistical significance of predefined pathways belonging to MSigDB collection in a colon cancer data set by using two known and statistically well founded methods. Our study highlights the importance of using gene sets for understanding the main biological processes and pathways involved in colorectal cancer. Our analysis shows that many of the genes involved in these pathways, although obtained in studies on different cancer diseases, are strongly associated to colorectal tumorigenesis.

8.1 Introduction

Gene expression profiling has become a mainstay in the current research in applied genomics [1]. Current clinical practice consists in collecting specimens of tissues in two different phenotypical conditions, such as diseased patients and healthy controls, and identifying the main pathways and biological processes involved in the analyzed pathology. Such processes are coded through lists of genes defined on the basis of a-priori biological knowledge or experimentally. In the first case, such lists are composed of those genes which cooperate or are co-expressed in a particular cellular mechanism or function [3, 4, 5]. In the second case, the gene set represents the signature (response) of cells (system) to a given stimulus [6]. The focus of the problem is to identify biological processes, cellular functions and pathways

Angela Distaso
ISSIA-CNR, Via Amendola 122/D-I, 70126 Bari, Italy.

* Corresponding author

perturbed in the phenotypic conditions by analyzing genes co-expressed in a given pathway as a whole, taking into account the possible interactions among them and, more important, the correlation of their expression with the phenotypical conditions [6, 9].

In this paper we describe the results obtained by applying this approach to a data set composed of gene expression profiles relative to a case-control study of patients affected by colon cancer. Two well known methods recently proposed for finding deregulated pathways were applied. GSEA (Gene Set Enrichment Analysis) [7] finds perturbed pathways comparing the rank of genes in the data set with the ones belonging to the given pathway. To this end a Kolmogorov-Smirnov like statistic is used for assessing the statistical significance of the deregulation. GLAPA (Gene List Analysis with Prediction Accuracy) [8] uses the prediction accuracy of the phenotypic status of the patients for finding the pathways involved in the pathology. Both use non parametric permutation tests [12] and false discovery rate (FDR) [2] for assessing the statistical significance of the estimates. The database of gene sets we use in this study is the Molecular Signatures Database (MSigDB) [7]. This is a collection of 1687 curated gene sets with sizes ranging from 2 to 1594 genes, obtained from online pathway databases, publications in PubMed and expert knowledge.

8.2 Materials and Methods

8.2.1 Data set description

The gene expression profile data set was collected in Casa Sollievo della Sofferenza Hospital, Foggia - Italy [11]. The data set is made up of 22 normal and 25 tumor specimens of patients affected by colon cancer, profiled using the Affymetrix (Santa Clara, CA) HGU133A GeneChip (22283 probe-sets).

8.2.2 GLAPA: Gene List Analysis with Prediction Accuracy

GLAPA uses a novel approach for finding deregulated gene sets [8]. As a measure of relevance or deregulation of a given gene set, the prediction accuracy of the phenotype is used. The rationale is that a functional category coded through a list of genes is perturbed in a particular disease if it is possible to correctly predict the occurrence of the pathology in new subjects on the basis of the expression levels of those genes only. In other words, a functional category is informative for or is deregulated in a disease if the expression levels of the genes involved in the category are useful for training classifiers able to generalize, that is, able to correctly predict the status of new subjects [26]. So, generalization ability of predictors trained by using the expression levels of the genes co-operating in a given cellular mechanism or function can be seen as a measure of the relevance of the function in the pathology at hand. With the aim of estimating the relevance of a given pathway L we compute the prediction error e_L of a linear Regularized Least Squares (RLS) classifier [50]. We trained the classifier by using the genes in the gene set with the strategy described in [8]. In particular we evaluated the error rate associated to the gene set performing the leave-k-out cross validation (LKOCV) strategy. The statistical significance of the measured accuracy is assessed against a couple of null hypothesis by using two independent permutation tests [12]. The first one (T1) aims at measuring how e_L is due to the actual correlation between the genes in L and the phenotype and how it is due by chance. To this end, we estimate the empirical probability density function of e_L under the null hypothesis H_0^y in which the genotype and the phenotype are supposed to be independent random variables. The nominal P-value p_y relative to e_L is so given by the percentage of random errors smaller than e_L. The second permutation test (T2) aims at evaluating how e_L is dependent on the n genes cooperating in the biological function coded by the list L and how it depends only on the size of the list. In particular, in this test we assess if lists of the same size as L, composed of genes randomly selected from the ones present on the microarray, produce error rates smaller than e_L. The nominal P-value p_n relative to e_L is estimated as the percentage of random errors smaller than e_L. Moreover to account for multiple hypothesis testing, an estimate of the False Discovery Rate (FDR) [2] is computed. FDR is defined by the proportion of false hypothesis findings over the amount of alternative hypotheses accepted at a given level of statistical significance. The same procedure is adopted for computing an estimate of FDR_y relative to p_y and an estimate FDR_n relative to p_n. Another parameter that we have considered in evaluating the tests is the power (π_y and π_n), defined as the probability of accepting the alternative hypothesis H_1 when it is true.

8.2.3 GSEA: Gene Set Enrichment Analysis

GSEA [7] provides a statistical method for assessing the significance of pre-defined gene-sets starting from a microarray experiment. Given a gene expression dataset, the genes are ordered in a ranked list S according to their differential expression between the two classes. GSEA provides a score which measures the degree of enrichment of a given gene-set L at the extremes (top or bottom) of the rank-ordered list S. GSEA is based on a maximum deviation statistic of two distribution functions, similarly to the Kolmogorov-Smirnov test that is used to estimate the difference between the distributions. In fact, the score is calculated by walking down the list S, increasing a running-sum statistic when a gene in the gene-set is encountered, and decreasing it when genes not belonging to the gene-set are encountered. The magnitude of the increment depends on the correlation of the gene with the phenotype. The enrichment score (ES) is the maximum deviation from zero in the walk. The gene-sets related to the phenotypic distinction will tend to show high values of the

ES. The significance of the ES is assessed by permutation testing: the observed ES is compared with the distribution of enrichment scores under the null hypothesis that the genotype and the phenotype are supposed to be independent random variables. The nominal p-value is given by the percentage of random enrichment scores greater than the observed value of ES. This procedure is similar to the one performed in the T1 permutation test of GLAPA.

Note that, with the aim of comparing analysis results across gene sets, the primary statistic suggested by the authors of GSEA, is the normalized enriched scores (NES). In fact by normalizing the ES, GSEA takes into account the differences in gene set size and correlations between gene sets and the expression data sets. NES is based on the gene sets enrichment scores for all dataset permutations. Hence in our experiment we refer to NES for examining the relevance of the gene sets.

8.3 Results and Discussion

8.3.1 Statistical analysis

The deregulation of the whole collection of gene sets belonging to MSigDB was measured applying GSEA and GLAPA tools on our colon cancer data set independently. The GSEA software parameters were set to their default values. The statistical significance of normalized enrichment score (NES) associated to each gene set was assessed through a non parametric permutation test in which 1000 random permutations of the phenotypic labels were carried out. GSEA found 915 gene sets up-regulated in tumor and 769 up-regulated in normal specimens. Among these, only 399 gene sets up-regulated in tumor and 3 up-regulated in normal were found statistically significant with $\widehat{FDR}_y \leq 25\%$.

For measuring the deregulations of each gene set L with GLAPA, we measured the prediction error of the phenotype e_L associated to L. To this end, for each gene set, 1000 cross validations of the data set were carried out. In each cross validation, we used 30 examples for training and the remaining 17 for testing RLS classifiers with linear kernel. We found 1381 pathways with an error rate $e_L \leq 25\%$. For assessing the statistical significance of e_L with the permutation test T1, 1000 random permutations of the phenotypic labels were performed. This permutation test revealed 690 statistically significant gene sets ($p_y \leq 0.01$, $\widehat{FDR}_y \leq 0.024$) having error rates $e_L \leq 17\%$. In order to determine if the deregulation of a particular pathway was due to the identity of the genes cooperating in the given pathway, or simply to the number of genes present in the gene set, the permutation test T2 was carried out. Specifically, 1000 gene sets were generated composed of n probes randomly drawn from the ones available on the microarray. The error rate associated to each random gene set was evaluated performing 200 cross validations and compared with the error rate e_L. Such analysis revealed 58 pathways ($p_n \leq 0.02$, $\widehat{FDR}_n \leq 0.25$) having an error rate $e_L \leq 11\%$ ($p_y \leq 0.010, \widehat{\text{FDR}}_y \leq 0.024$). The table 8.1 shows the 21 statistically significant pathways found deregulated by both methods.

8.3.2 Biological and functional analysis

We analyzed in depth some gene sets found deregulated with high statistical significance in the current experimental conditions for finding biological confirmations of their involvement in the pathology. In particular, we studied 2 pathways found perturbed by both methods: *ADIPOCYTE_BRCA_UP* and *CELL_CYCLE_CHECKPOINT*. Moreover, for showing the ability of GLAPA of determining new pathways not found by GSEA, we studied a third gene set, *HDACI_COLON_SUL12HRS_UP*, found deregulated by GLAPA only.

Firstly, we analyzed *ADIPOCYTE_BRCA_UP* gene set that is composed of ten genes. At a first analysis it seems no related to colon cancer, because this experimentally determined gene set results upregulated in breast cancer cells (MCF-7) treated with adipocyte-conditioned growth media [27]. However, an analysis in detail of the genes co-expressed shows a strong correlation of this pathway with colon cancer. **ATF3** (activating transcription factor 3; Location: 1q32.3) is an eukaryotic transcription factor that is upregulated transcriptionally during cellular responses to a variety of stresses, in particular DNA damage [28]. Moreover dysfunction of ATF3 impairs the p53-mediated cellular response to DNA damage, allowing cells to be readily transformed by oncogenes. This notion is consistent with the observation of downregulated ATF3 expression in most human cancers [29], [30]. Furthermore Lee et al. have reported that ATF3 may play a pivotal role in DIM(3,3'-diindolylmethane)-induced NAG-1(Nonsteroidal anti-inflammatory drug-activated gene-1) expression in human colorectal cancer cells [31].

The second gene analyzed was **IGF2** (insulin-like growth factor 2; Location: 11p15.5) that plays a critical role in the regulation of cell growth and transformation. IGF-I and IGF-II inhibit apoptosis, promote tumor growth, and induce transformation and metastasis in many types of malignancies. The gastrointestinal system may be one of the major targets of IGF action and there is increasing evidence that alterations in IGF signaling are involved in the neoplastic transformation and progression of colorectal carcinoma. A significant overexpression of IGF-II mRNA and protein levels has been reported in 30-40% of colorectal carcinoma patients [32].

Another gene analyzed was **MMP1** (matrix metallopeptidase 1 ; Location: 11q22.3). It is a component of matrix metalloproteinases (MMPs) that collectively degrade most of the components of the extracellular matrix (ECM), so these MMPs are believed to contribute to the proliferation, invasion and metastasis of tumor cells by eliminating the surrounding ECM barrier [33] [34]. In addition, MMPs are required for tumor-induced angiogenesis, loss of cell adhesion, tumor growth and apoptosis [33]. Numerous MMPs, including MMP1, MMP3 and MMP7, are overexpressed in tumors and they have been associated also with development of colorectal cancer [34] [35].

Another gene belonging to ADIPOCYTE_BRCA_UP gene set was **NF-kB** (nuclear factor of kappa light polypeptide gene enhancer in B-cells 1; Location: 4q24) that is a generic name for a transcription-factor system that is involved in the regulation of cell proliferation, development, and apoptosis. The analysis of the expression of NFkB in various colorectal carcinoma cell lines shows that the inactive cytoplasmic NFkB form is evidently up-regulated in the tumor epithelium, especially in the metastatic cases, as compared to normal tissue.

The transcription factor **SOX9** (sex determining region Y-box 9; Location: 17q24.3-q25.1) is another gene co-expressed in this pathway. It was found inducing apoptosis in a human colon carcinoma cell line [36].

Altered patterns of **STC1** (stanniocalcin 1; Location: 8p21-p11.2) expression have a role in human cancer development. Hypoxia can stimulate STC1 gene expression in various human cancer cell lines, including those derived from colon carcinomas [37].

Finally we have analyzed **TNFAIP3** (tumor necrosis factor, alpha-induced protein 3; Location:6q23) gene that may act as a tumor suppressor gene inhibiting NFkB activity and tumor necrosis factor (TNF)-mediated programmed cell death.

Secondly, we considered *CELL_CYCLE_CHECKPOINT* gene set, belonging to Gene Ontology (GO) data base and composed of twenty six genes. Cell cycle checkpoints are essential in eukaryotes for ensuring high fidelity transmission of genetic information from one generation to the next. They include DNA damage checkpoints, DNA replication checkpoints, spindle assembly checkpoints, and cytokinesis checkpoints. Also in this case we give a short description of the single genes belonging to this gene set, underlining their importance in oncogenesis.

ABL1 (v-abl Abelson murine leukemia viral oncogene homolog 1; Location:9q34.1) proto-oncogene encodes a protein tyrosine kinase that has been implicated in processes of cell differentiation, cell division, cell adhesion, death, and stress response. Several findings suggest that the 9q34 region was altered in some cases of sporadic colorectal carcinomas [38].

The protein encoded by **ATM** (ataxia telangiectasia mutated; Location:11q22-q23) gene is an important cell cycle checkpoint kinase that regulates a wide variety of downstream proteins. ATM and the closely related kinase ATR are thought to be master controllers of cell cycle checkpoint signaling pathways. Frequent allelic imbalances at the ATM locus have been reported in colorectal cancer and some findings led us to hypothesize that loss of expression of this gene may have a role in the early stage of colorectal cancer development and it may be related to advanced tumor stage and poorer patient survival [39].

Another interesting gene analyzed was **BRCA1** (breast cancer 1; Location: 17q21) that encodes a nuclear phosphoprotein that plays a role in maintaining genomic stability and acts as a tumor suppressor. Defects in BRCA1 are a cause of genetic susceptibility to breast cancer. Moreover, BRCA1 mutation carriers have a 4-fold increased risk of colon cancer, and loss of heterozygosity at the BRCA1 gene locus was shown to be associated with shorter survival in colorectal cancer. Recent evidences show that the expression of ATM and BRCA1 is a prognostic marker in colorectal cancer [39].

The protein kinase Chk1 encoded by **CHEK1** (checkpoint homolog; Location: 11q24-q24) gene is essential in human cells for cell cycle arrest in response to DNA damage, and has been shown to play an important role in the G2/M checkpoint. Some results suggest that the Chk1 gene is a target of genomic instability in microsatellite instability (MSI)-positive colorectal cancers and that the Chk1 frameshift mutations might be involved in colorectal tumorigenesis through a defect in response to DNA damage [40].

Also **CHEK2** (checkpoint homolog; Location: 22q12.1) is a cell cycle checkpoint regulator and it was considered a putative tumor suppressor stabilizing the tumor suppressor protein p53 and leading to cell cycle arrest in G1. Recently, a functionally defective CHEK2 variant I157T has been proposed to be associated with an increased risk of colorectal cancer in a large population based study including a significant number of familial and sporadic colorectal cancer cases [41].

Also analysis of **GADD45** (growth arrest and DNA-damage-inducible, alpha; Location: 1p31.2-p31.1) was interesting. It is a growth arrest-associated gene that is induced in response to DNA damage. Alterations of the expressions of GADD45, ZBRK1, and BRCA1 genes have been reported in colon carcinomas [42].

Another important finding was **MAD2L1** (MAD2 mitotic arrest deficient-like 1; Location: 4q27) expression in colorectal cancer that was higher than that in the corresponding normal tissue. The expression of Mad2 in colorectal cancer was related with histological differentiation and lymph node metastasis. Li et al. 2004, showed that Mad2 protein overexpressed in cancer tissue might be good marker for predicting histological differentiation and prognosis of colorectal cancer [43].

Another important component of this pathway was **MRE11A** (MRE11 meiotic recombination 11 homolog A Location: 11q21) gene which encodes a nuclear protein involved in homologous recombination, telomere length maintenance, and DNA double-strand break repair. MRE11 may be considered as a new target in the mismatch repair deficient tumorigenesis with a role in colorectal carcinogenesis [44].

Also mutations of the mismatch repair gene, **MSH3** (mutS homolog 3 ; Location: 5q11-q12), might play a role in the progression of mismatch repair deficient tumors by increasing instability. Common polymorphisms in MSH3 may increase the risk of colorectal cancer, especially proximal colon cancer [45].

Also **NBS1** (nibrin; Location: 8q21), which plays a critical role in the cellular response to DNA damage and the maintenance of chromosome integrity, could be a tumor suppressor gene involved in proximal colorectal cancer [46].

Another important gene analyzed was **RAD17** (RAD17 homolog; Location: 5q13) that is a cell cycle checkpoint gene required for cell cycle arrest and DNA damage repair in response to DNA damage. Rad17 is overexpressed in various cancer cell lines and in colon carcinoma. Its chromosomal localization suggests that a variety of human cancers would have a deletion in this gene [47].

RPA1 (replication protein A1; Location: 17p13.3) and **RPA2** (replication protein A2; Location: 1p35) are required for stabilization of single-stranded DNA at early and later stages of DNA replication being thus critical for eukaryotic DNA replication. Experimental studies in colon cancer cell lines have shown that RPA proteins may be useful prognostic indicators in colon cancer patients and attractive therapeutic targets for regulation by tumor suppressors [48].

Another important gene belonging to this gene set was **TP53** (tumor protein p53; Location: 17p13.1) gene, that plays an essential role in the regulation of cell cycle, specifically in the transition from G0 to G1. p53 activates expression of downstream genes that inhibit growth and/or invasion, and thus function as a tumor suppressor. Over 8000 mutations of this gene have been identified among tumor types. The G→A transition in codon 175 of p53 gene may be useful as a potential marker of colorectal cancer progression and in evaluating the margins of surgical resection [49].

Finally, we focused on the *HDACI_COLON_SUL12HRS_UP* gene set that was found deregulated by GLAPA software only. Our aim was to find biological confirmation of its statistical significance. This gene set, composed of twenty six genes, seems to be specific for colorectal cancer. It was obtained experimentally by SW620 colonic epithelial cells as described in [13]. **ANXA2** (annexin A2; Location: 15q21-q22) and **ANXA5** (annexin A5; Location: 4q28-q32) genes encode two members of the annexin family, a calcium-dependent phospholipid-binding protein family that plays a role in the regulation of cellular growth and in signal transduction pathways. It has been suggested that annexin 2 and annexin 5 are involved in cell proliferation/differentiation and the pathogenesis of carcinoma. Their overexpression has been reported in various carcinomas including colon malignant tumors, and some findings suggest that they may be related to the progression and metastatic spread of colorectal carcinoma [14].

The second gene analyzed was **API5** (apoptosis inhibitor 5; Location: 11p11.2) gene that is an inhibitor of apoptosis. Many growth factors and cytokines act as cellular survival factors by preventing programmed cell death. Its expression is often upregulated in tumor cells, particularly in metastatic cells; so inhibition of Api5 function might offer a possible mechanism for antitumor exploitation [15].

Another gene belonging to this list is decay-accelerating factor (DAF) **CD55** (decay accelerating factor for complement; Location: 1q32) that is a membrane glycoprotein that regulates complement activation. The expression of DAF is enhanced in colorectal cancer cells and the colonic epithelium of ulcerative colitis in relation to the degree of mucosal inflammation [16] [17].

The **CDH1** (cadherin 1, type 1, E-cadherin (epithelial); Location: 16q22.1) gene belongs to the cadherin superfamily. The encoded protein is a calcium dependent cell-cell adhesion glycoprotein. Mutations in this gene are correlated with gastric, breast, colorectal, thyroid and ovarian cancer. The examination of E-cadherin expression and distribution in colorectal tumors can be extremely valuable in predicting disease recurrence [10].

Another important gene analyzed was **GSR** (glutathione reductase; Location: 8p21.1). The gastrointestinal tract is particularly susceptible to reactive oxygen species attack which lead to carcinogenesis. An important role in defense strategy against reactive oxygen species is played by antioxidants. In fact, superoxide dismutase, glutathione peroxidase and reductase are in general over-expressed in colorectal tumor [18].

HSP90AA1 (heat shock protein 90kDa alpha (cytosolic), class A member 1; Location: 14q32.33) gene is a member of molecular chaperones, specifically of the heat shock protein 90 (Hsp90) family. HSP90 was low or non-detectable in normal colon tissues while high levels of HSP90 expression were observed in human colon cancer tissues, confirming the role of HSP90 as a potential marker for malignant colon cancer [19].

The protein encoded by **MYC** (v-myc myelocytomatosis viral oncogene homolog (avian); Location: 8q24.21) gene is a multifunction, nuclear phosphoprotein that plays a role in cell cycle progression, apoptosis and cellular transformation. It functions as a transcription factor that regulates transcription of specific target genes. Mutations, overexpression, rearrangement and translocation of this gene have been associated with a variety of tumors. MYC protects from p53-mediated apoptosis. Some findings indicate that failure of the normal apoptotic process together with de-regulation of MYC proto-oncogene might promote the development of colorectal tumors and its overexpression is observed in most colorectal cancers [20] [21].

An essential requirement for the development, progression and metastasis of malignant tumors is angiogenesis. **VEGF** (vascular endothelial growth factor; Location: 6p12) gene plays an essential role in the development of angiogenesis of numerous solid malignancies, including colon cancer. This gene is a member of the PDGF/VEGF growth factor family and acts on endothelial cells mediating increased vascular permeability, inducing angiogenesis, vasculogenesis and endothelial cell growth, promoting cell migration, and inhibiting apoptosis. VEGF is associated with the development and prognosis of colorectal cancer, but its relation with degree of differentiation remains to be studied [22]; likely VEGF functions as regulators of colon cancer cell invasion. VEGF expression is induced in colon and other cancer cells as a result of hypoxia and multiple genetic alterations. However it is evident that there is an association between VEGF expression, p53 status and angiogenesis, suggesting that mutant p53 plays a central role in promoting angiogenesis in colon cancer progression [23].

8.4 Conclusions

In this paper we have described the biological and functional relevance in colon cancer of pathways found deregulated in a gene expression profile data set relative to a case-control study of patients affected by this pathology [11]. GLAPA and GSEA methods were applied for measuring deregulation of pathways and for assessing their statistical significance [7, 8]. The pathway database used in this study is a curated collection of 1687 gene sets obtained from different sources [7]. Other studies have pointed out the fundamental role of pathways in studying onset and progression of tumors [6, 24]. Indeed cancer is a heterogeneous disease caused by a complex of altered processes that could be grouped according to the six hallmarks of cancer (self-sufficiency in growth signals, insensitivity to anti-growth signals, evasion of apoptosis, limitless replicative potential, sustained angiogenesis, tissue invasion and metastasis) [25]. Our study highlights that pathway approach to the investigation of complex diseases allows to get a well comprehensive picture of altered biological processes in cancer pathology.

Pathway	Size	GLAPA							GSEA		
		e_L	p_y	FDR_y	π_y	p_n	FDR_n	π_n	NES	p_y	FDR_y
ADIPOCYTE BRCA UP	17	0.09	0.003	0.02	0.97	0.008	0.12	0.68	1.7	0.016	0.04
ADIP DIFF CLUSTER3	68	0.08	0.001	0.02	0.99	0.002	0.04	0.78	1.5	0.032	0.15
AS3 HEK293 DN	20	0.10	0.003	0.02	0.96	0.008	0.12	0.72	1.7	0.006	0.05
BLEO HUMAN LYMPH HIGH 4HRS UP	33	0.09	0.001	0.02	0.97	0.007	0.12	0.73	1.4	0.129	0.22
BLEO MOUSE LYMPH HIGH 24HRS DN	79	0.11	0.002	0.02	0.97	0.015	0.21	0.61	1.6	0.052	0.08
CANCER UNDIFFERENTIATED META UP	96	0.09	0.002	0.02	0.98	0.008	0.12	0.69	1.9	0.001	0.01
CARBON FIXATION	37	0.09	0.009	0.02	0.84	0.008	0.12	0.74	1.59	0.029	0.10
CELL CYCLE CHECKPOINT	47	0.11	0.004	0.02	0.97	0.02	0.24	0.60	1.81	0.008	0.03
CIS RESIST LUNG UP	19	0.11	0.004	0.02	0.95	0.02	0.24	0.66	1.55	0.049	0.12
CMV HCMV TIMECOURSE 14HRS UP	77	0.11	0.001	0.02	0.96	0.02	0.24	0.64	1.61	0.021	0.09
CROONQUIST IL6 RAS DN	37	0.10	0.001	0.02	0.97	0.02	0.24	0.66	1.71	0.018	0.05
FATTY ACID SYNTHESIS	26	0.11	0.003	0.02	0.97	0.02	0.24	0.62	1.54	0.048	0.13
HG PROGERIA DN	39	0.11	0.002	0.02	0.97	0.02	0.24	0.67	1.84	0.004	0.02
HIFPATHWAY	31	0.10	0.002	0.02	0.97	0.01	0.13	0.69	1.69	0.013	0.06
JAIN NEMO DIFF	125	0.09	0.001	0.02	0.99	0.01	0.05	0.72	1.86	0.002	0.02
KLEIN PEL UP	102	0.09	0.002	0.02	0.99	0.01	0.09	0.71	1.73	0.006	0.05
NEMETH TNF DN	53	0.09	0.002	0.02	0.98	0.01	0.04	0.74	1.77	0.002	0.04
PENTOSE PHOSPHATE PATHWAY	35	0.07	0.002	0.02	0.93	0.001	0.001	0.85	1.61	0.025	0.09
PURINE METABOLISM	191	0.10	0.002	0.02	0.98	0.015	0.212	0.63	1.85	0.002	0.021
SA G2 AND M PHASES	16	0.11	0.003	0.02	0.96	0.016	0.225	0.65	1.70	0.025	0.056
ZHAN MMPC SIM BC AND MM	88	0.10	0.002	0.02	0.98	0.021	0.243	0.65	1.54	0.035	0.126

Table 8.1 *Pathways of MSigDB database found deregulated by both GLAPA and GSEA methods in our colon cancer gene expression data set. For each pathway we report the name, the number of probes (size) and the most relevant statistical parameters as measured by both methods.*

References

1. Schena M, Shalon D, Davis RW, Brown, PO. Quantitative monitoring of gene-expression patterns with a complementary-DNA microarray. Science, (270), 467-470 (1995)
2. Storey JD, Tibshirani R. Statistical significance for genomwide studies. Proc. Natl. Acad. Sci., (100), 9440-9445 (2003)
3. Ashburner M, Ball CA, Blake JA, Botstein D, Butler H, Cherry JM, Davis AP, Dolinski K, Dwight SS, Eppig JT, Harris MA, Hill DP, Issel-Tarver L, Kasarskis A, Lewis S, Matese JC, Richardson JE, Ringwald M, Rubin GM, Sherlock G. Gene ontology: tool for the unification of biology. The Gene Ontology Consortium. Nat Genet, (25), 25-29 (2000)
4. Kanehisa M, Goto S, Kawashima S, Nakaya A. The KEGG databases at GenomeNet. Nucleic Acids Res, (30), 42-46 (2002)
5. Khatri P, Draghici S, Ostermeier GC, Krawetz SA. Profiling Gene Expression Using Onto-Express. Genomics, 79(2), 266-270 (2002)
6. Bild AH, Yao G, Chang JT, Wang Q, Potti A, Chasse D, Joshi MB, Harpole D, Lancaster JM, Berchuck A, Olson JA, Marks JR, Dressman HK, West M, Nevins JR. Oncogenic pathway signatures in human cancers as a guide to targeted therapies. Nature, 19(439), 353-357 (2006)
7. Subramanian A, Tamayo P, Mootha VK, Mukherjee S, Ebert BL, Gillette MA, Paulovich A, Pomeroy SL, Golub TR, Lander ES, Mesirov JP. Gene set enrichment analysis: A knowledge-based approach for interpreting genome-wide expression profiles. Proc. Natl. Acad. Sci., (102), 15545-15550 (2005)
8. Maglietta R, Piepoli A, Catalano D, Licciulli F, Carella M, Liuni S, Pesole G, Perri F, Ancona N. Statistical assessment of functional categories of genes deregulated in pathological conditions by using microarray data. Bioinformatics, 23(16), 2063-2072 (2007)
9. Creighton CJ. Multiple oncogenis Pathway Signatures Show Coordinate Expression Patterns in Human Prostate Tumors. PLoS ONE, 3(3), e1816 (2008)
10. Elzagheid A, Algars A, Bendardaf R, Lamlum H, Ristamaki R, Collan Y, Syrjanen K, Pyrhonen S. E-cadherin expression pattern in primary colorectal carcinomas and their metastases refl ects disease outcome. World J Gastroenterol., 12(27), 4304-4309 (2006)
11. Ancona N, Maglietta R, Piepoli A, D'Addabbo A, Cotugno R, Savino M, Liuni S, Carella M, Pesole G, Perri F. On the statistical assessment of classifiers using DNA microarray dataBMC Bioinformatics, 387(7) (2006)
12. Good P. Permutation tests: a practical guide to resampling methods for testing hypotheses. Springer Verlag, New York Inc (1994)
13. Mariadason JM, Corner GA, Augenlicht LH. Genetic reprogramming in pathways of colonic cell maturation induced by short chain fatty acids: comparison with trichostatin A, sulindac, and curcumin and implications for chemoprevention of colon cancer. Cancer Res., 60(16), 4561-4572 (2000)
14. Emoto K, Yamada Y, Sawada H, Fujimoto H, Ueno M, Takayama T, Kamada K, Naito A, Hirao S, Nakajima S. Annexin II Overexpression Correlates with Stromal Tenascin-C Overexpression. A Prognostic Marker in Colorectal Carcinoma. Cancer, (92), 1419-1426 (2001)
15. Morris EJ, Michaud WA, Ji JY, Moon NS, Rocco JW, Dyson NJ. Functional Identification of Api5 as a Suppressor of E2F-Dependent Apoptosis In Vivo. PLoS Genetics, 2(11), 1834-1848 (2006)
16. Okazazi H, Mizuno M, Nasu J, Makidono C, Hiraoka S,Yamamoto K, Okada H, Fujita T, Tsuji T, Shiratori Y. Difference in Ulex europaeus agglutinin I-binding activity of decay-accelerating factor detected in the stools of patients with colorectal cancer and ulcerative colitis. J Lab Clin Med, 143(3), 169-174 (2004)
17. Durrant LG, Chapman MA, Buckley DJ, Spendlove I, Robins RA, Armitage NC. Enhanced expression of the complement regulatory protein CD55 predicts a poor prognosis in colorectal cancer patients. Cancer Immunol. Immunother., (52)(2003)
18. Skrzydlewska E, Kozuszko B, Sulkowska M, Bogdan Z, Kozlowski M, Snarska J, Puchalski Z, Sulkowski S, Skrzydlewski Z. Antioxidant potential in esophageal, stomach and colorectal cancers. Hepatogastroenterology, 50(49), 126-131 (2003)

19. Park KA, Byun HS, Won M, Yang KJ, Shin S, Piao L, Kim JM, Yoon WH, Junn E, Park J, Seok JH, Hur GM. Sustained activation of protein kinase C downregulates nuclear factor-kB signaling by dissociation of IKK-g and Hsp90 complex in human colonic epithelial cells. Carcinogenesis, 28(1), 71-80 (2007)
20. Greco C, Alvino S, Buglioni S, Assisi D, Lapenta R, Grassi A, Stigliano V, Mottolese M, Casale V. Activation of c-MYC and c-MYB proto-oncogenes is associated with decreased apoptosis in tumor colon progression. Anticancer Res., 21(5), 3185-3192 (2001)
21. Seidler HBK, Utsuyama M, Nagaoka S, Takemura T, Kitagawa M, Hirokawa K. Expression level of Wnt signaling components possibly influences the biological behavior of colorectal cancer in different age groups. Experimental and Molecular Pathology, (76), 224-233 (2004)
22. Han J, Xia C, Gao J, Xing C, Yang X, Tang X, Qiu F, Du Y. Expression of vascular endothelial growth factor in colorectal cancer and its clinical significance. Zhonghua Yi Xue Za Zhi, 82(7), 481-483 (2002)
23. Faviana P, Boldrini L, Spisni R, Berti P, Galleri D, Biondi R, Camacci T, Materazzi G, Pingitore R, Miccoli P, Fontanini G. Neoangiogenesis in colon cancer: correlation between vascular density, vascular endothelial growth factor (VEGF) and p53 protein expression. Oncol Rep., 9(3), 617-620 (2002)
24. Edelman EJ, Guinney J, Chi JT, Febbo PG, Mukherjee S. Modeling cancer progression via pathway dependencies. PLoS Computational Biology, 4(2), e28 (2008)
25. Hanahan D, Weinberg RA. The Hallmarks of Cancer. Cell, (100), 57-70 (2000)
26. Vapnik V. The Nature of Statistical Learning Theory. New York, Inc.: Springer Verlag 1995.
27. Iyengar P, Combs TP, Shah SJ, Gouon-Evans V, Pollard JW, Albanese C, Flanagan L, Tenniswood MP, Guha C, Lisanti MP, Pestell RG, Scherer PE. Adipocyte-secreted factors synergistically promote mammary tumorigenesis through induction of anti-apoptotic transcriptional programs and proto-oncogene stabilization. Oncogene, 22(41), 6408-6423 (2003)
28. Wang A, Arantes S, Conti C, McArthur M, Aldaz CM, MacLeod MC. Epidermal hyperplasia and oral carcinoma in mice overexpressing the transcription factor ATF3 in basal epithelial cells. Md. Carcinog., 46(6), 476-487 (2007)
29. Yan C, Boyd DD. ATF3 regulates the stability of p53: a link to cancer. Cell Cycle, 5(9), 926-929 (2006)
30. Zhang C, Gao C, Kawauchi J, Hashimoto Y, Tsuchida N, Kitajima S. Transcriptional activation of the human stress-inducible transcriptional repressor ATF3 gene promoter by p53. Biochem. Biophys. Res. Commun., 297, 1302-1310 (2002)
31. Lee SH, Kim JS, Yamaguchi K, Eling TE, Baek SJ. Indole-3-carbinol and 3,3'-diindolylmethane induce expression of NAG-1 in a p53-independent manner. Biochem. Biophys. Res. Commun., 328(1), 63-69 (2005)
32. Weber MM, Fottner C, Liu SB, Jung MC, Engelhardt D, Baretton GB. Overexpression of the Insulin-Like Growth Factor I Receptor in Human Colon Carcinomas. Cancer, 95(10), 2086-2095 (2002)
33. Seiki M. Membrane-type 1 matrix metalloproteinase: a key enzyme for tumor invasion. Cancer Letters, 2(194), 1-11 (2003)
34. Ye S. Polymorphism in matrix metalloproteinase gene promoters: implication in regulation of gene expression and susceptibility of various diseases. Matrix Biol., 19, 623-629 (2000)
35. Hewitt RE, Leach IH, Powe DG, Clark IM, Cawston TE, Turner DR. Distribution of collagenase and tissue inhibitor of metalloproteinases (TIMP) in colorectal tumours. Int.J.Cancer., 49, 666-672 (1991)
36. Jay P, Berta P, Blache P. Expression of the Carcinoembryonic Antigen Gene Is Inhibited by SOX9 in Human Colon Carcinoma Cells. Cancer Res., 65(6), 2193-2198 (2005)
37. Yeung HY, Lai KP, Chan HY, Mak NK, Wagner GF, Wong CKC. Hypoxia-Inducible Factor-1-Mediated Activation of Stanniocalcin-1 in Human Cancer Cells. Endocrinology, 146(11), 4951-4960 (2005)
38. Bartos JD, Gaile DP, McQuaidb DE, Conroy JM, Darbary H, Nowak NJ, Block A, Petrelli NJ, Mittelman A, Stoler DL, Anderson GR. aCGH local copy number aberrations associated with overall copy number genomic instability in colorectal cancer: Coordinate involvement of the regions including BCR and ABL. Mutation Research, 615, 1-11 (2007)

39. Grabsch H, Dattani M, Barker L, Maughan N, Maude K, Hansen O, Gabbert HE, Quirke P, Mueller W. Expression of DNA Double-Strand Break Repair Proteins ATM and BRCA1 Predicts Survival in Colorectal Cancer. Clin.Cancer Res., 12(5), 1494-1500 (2006)
40. Kim CJ, Lee JH, Song JW, Cho YG, Kim SY, Nam SW, Yoo NJ, Park WS, Lee JY. Chk1 frameshift mutation in sporadic and hereditary non-polyposis colorectal cancers with microsatellite instability. EJSO, 33, 580-585 (2007)
41. Kilpivaara O, Alhopuro P, Vahteristo P, Aaltonen LA, Nevanlinna HJ. CHEK2 I157T associates with familial and sporadic colorectal cancer. Med.Genet., 43(7), e:34 (2006)
42. Garcia V, Garcia JM, Pena C, Silva J, Dominguez G, Rodriguez R, Maximiano C, Espinosa R, Espana P, Bonilla F. The GADD45, ZBRK1 and BRCA1 pathway: quantitative analysis of mRNA expression in colon carcinomas. J.Pathol., 206(1), 92-99 (2005)
43. Li GQ, Zhang HF. Mad2 and p27 expression profiles in colorectal cancer and its clinical significance. World J Gastroenterol., 10(21), 3218-3220 (2004)
44. Giannini G, Rinaldi C, Ristori E, Ambrosini MI, Cerignoli F, Viel A, Bidoli E, Berni S, D'Amati G, Scambia G, Frati L, Screpanti I, Gulino A. Mutations of an intronic repeat induce impaired MRE11 expression in primary human cancer with microsatellite instability. Oncogene, 23(15), 2640-2647 (2004)
45. Berndt SI, Platz EA, Fallin MD, Thuita LW, Hoffman SC, Helzlsouer KJ. Mismatch repair polymorphisms and the risk of colorectal cancer. Int.J.Cancer, 120(7), 1548-1554 (2007)
46. Uhrhammer N, Bay JO, Gosse-Brun S, Kwiatkowski F, Rio P, Daver A, Bignon YJ. Allelic imbalance at NBS1 is frequent in both proximal and distal colorectal carcinoma. Oncol.Rep., 7(2), 427-431 (2000)
47. Bao S, Chang MS, Auclair D, Sun Y, Wang Y, Wong WK, Zhang J, Liu Y, Qian X, Sutherland R, Magi-Galluzi C, Weisberg E, Cheng EYS, Hao L, Sasaki H, Campbell MS, Kraeft SK, Loda M, Lo KM, Chen LB. HRad17, a Human Homologue of the Schizosaccharomyces pombe Checkpoint Gene rad17, Is Overexpressed in Colon Carcinoma. Cancer Res., 59, 2023-2028 (1999)
48. Givalos N, Gakiopoulou H, Skliri M, Bousboukea K, Konstantinidou AE, Korkolopoulou P, Lelouda M, Kouraklis G, Patsouris E, Karatzas G. Replication protein A is an independent prognostic indicator with potential therapeutic implications in colon cancer. Modern Pathology, 20, 159-166 (2007)
49. Krajewska WM, Stawinska M, Brys M, Mlynarski W, Witas H, Okruszek A, Kilianska ZM. Genotyping of p53 codon 175 in colorectal cancer. Med.Sci.Monit., 9(5), 228-231 (2003)
50. Rifkin R, Yeo G, Poggio T. Regularized least squares classification. Advances in Learning Theory: Methods, Model and Applications, NATO Science Series III: Computer and Systems Sciences, Edited by Suykens, Horvath, Basu, Micchelli, and Vandewalle, Amsterdam: IOS Press, 190, 131-153 (2003)

Chapter 9
A Genome-Wide Computational Study of Copy Number Variations: an Example on Ovarian Cancer

Anneleen Daemen, Olivier Gevaert, Karin Leunen, Vanessa Vanspauwen, Geneviève Michils, Eric Legius, Ignace Vergote, and Bart De Moor

Abstract *Motivation:* Knowledge about the molecular mechanisms involved in sporadic and hereditary ovarian tumorigenesis is lacking. Due to the hypothesis that BRCA related ovarian cancer follows distinct pathways in their carcinogenesis, array comparative genomic hybridization (array CGH) was performed in 8 sporadic and 5 BRCA1 mutated ovarian cancer patients to identify copy number variations.
Results: Chromosomal regions characterizing each group of sporadic and BRCA1 related ovarian cancer were gathered using recurrent hidden Markov Models (HMM). The differential regions were reduced to a subset of features for classification by integrating different univariate feature selection methods. Least Squares Support Vector Machines (LS-SVM), a supervised classification method, resulted in a leave-one-out accuracy of 84.6%, sensitivity of 100% and specificity of 75%.
Conclusion: The combination of recurrent HMMs for the detection of copy number alterations with LS-SVM classifiers offers a novel methodological approach for classification based on copy number alterations. Additionally, this approach limits the chromosomal regions that are necessary to distinguish sporadic from hereditary ovarian cancer.

9.1 Introduction

In cancers, many gains and losses of chromosomes and chromosomal segments have been described. These aberrations defined as regions of increased or decreased DNA copy number can be detected at high resolution using an array comparative genomic hybridization (array CGH) technology. This technique measures variations in DNA copy number within the entire genome of a disease sample compared to a normal sample [1]. This makes array CGH ideally suitable for a genome-wide identification

Anneleen Daemen
Department of Electrical Engineering (ESAT), Katholieke Universiteit Leuven, Leuven, Belgium

and localization of genetic alterations involved in human diseases. An overview of algorithms for array CGH data analysis is given in [2]. Segmentation approaches identify chromosomal regions of adjacent clones with the same mean log ratio. Disadvantages of these methods are that the segments that are gained or lost need to be determined in a further analysis and that results become unsatisfactory with high noise levels in the data. Therefore, segmentation and identification should be performed simultaneously because these two tasks can improve each other's performance. A popular method for combining them is the hidden Markov Model (HMM) with states defined as loss, neutral, one-gain and multiple-gain. Recently, this traditional procedure has been extended to a recurrent HMM in which a class of samples instead of individual samples is modeled by sharing information on copy number variations across multiple samples [4]. Here, we present a method to identify copy number alterations with the recurrent HMM which goes beyond the exploratory phase by using these alterations as features in a supervised classification setting and by validating these features biologically.

Because the exclusion of redundant and non-discriminatory features might avoid overfitting and identifies a smaller set of features able to distinguish good from bad, feature selection should be performed. To get rid of some of the arbitrariness with which a univariate feature selection method is chosen, different univariate test statistics were combined to suppress the false positive error rate [5]. For classification, we used the class of kernel methods which is powerful for pattern analysis. In recent years, these methods have become a standard tool in data analysis, computational statistics, and machine learning applications [6], [7]. Their rapid uptake in bioinformatics [8] is due to their reliability, accuracy and computational efficiency, which has been demonstrated in countless applications [9]. More specifically, as supervised classification algorithm we made use of the Least Squares Support Vector Machine (LS-SVM) which is an extension of the standard SVM and has been developed in our research group by Suykens *et al.* (1999), (2002) [10]-[11]. On high dimensional data, the LS-SVM is easier and faster to solve because the quadratic programming problem of the SVM is reduced to a set of linear equations.

We applied our method on ovarian cancer which is the fourth most common cause of cancer death and ranks as the most frequent cause of death from gynaecological malignancies among women in western countries [12]. In a total of 5-10% of epithelial ovarian carcinomas, a family history of breast and ovarian cancer is noted with germline mutations in the tumour suppressor genes BRCA1 or BRCA2 in most of them. A mutation of the BRCA1 gene cumulates the risk for ovarian carcinoma with 26-85% while a BRCA2 mutation increases the cumulative risk with 10% [14]. The knowledge of different copy number variations between both sporadic and hereditary groups may help to better understand tumorigenesis of these cancers. When applied to larger study groups, this method could result in a better comprehension of the different clinical behaviour of both groups, probably necessitating different treatment strategies.

The outline of this chapter is as follows. In section 9.2, we describe the data set and the array CGH technology used for the analysis as well as the recurrent HMM, the classifier and the feature selection method applied. In addition, the workflow of our proposed methodology is given in detail together with the functional annotation analysis to validate agreement of the selected chromosomal regions with biology. We determine the gene sets from the Molecular Signatures Database (MSigDB) [13] that are enriched in the identified regions of copy number alteration. We describe our results on ovarian cancer in Section 9.3 and conclude in Section 9.4.

9.2 Materials and Methods

9.2.1 Patients and Data

The data for this study were collected from patients treated for ovarian cancer at the University Hospital of Leuven, Belgium. A distinction could be made between patients with a sporadic tumour and carriers of a mutation in the tumour suppressor genes BRCA1 or BRCA2. Both genes are involved in DNA damage repair and transcriptional regulation [15]. All tumour samples were collected at the time of primary surgery. Only patients with similar clinical characteristics were retained: eight sporadic and five BRCA1 mutated ovarian cancer patients. One patient with BRCA2 was excluded and none of the patients out of the sporadic group had a positive family history of breast and/or ovarian cancer. Array comparative genomic hybridization was performed using a 1Mb array CGH platform, version CGH-SANGER 3K 7 developed by the Flanders Institute for Biotechnology (VIB), Department of Microarray Facility, Leuven, Belgium.

9.2.2 Array Comparative Genomic Hybridization

Array comparative genomic hybridization (array CGH) is a high-throughput technique for measuring DNA copy number variations (CNV) within the entire genome of a disease sample relative to a normal sample [1]. In an array CGH experiment, total genomic DNA from tumour and normal reference cell populations are isolated and subsequently labeled with different fluorescent dyes before being hybridized to several thousands of probes on a glass slide. This allows to calculate the log ratios of the fluorescence intensities of the tumour to that of the normal reference DNA. Because the reference cell population is normal, an increase or decrease in the log intensity ratio indicates a DNA copy number variation in the genome of the tumour cells such that negative log ratios correspond to deletions (losses), positive log ratios to gains or amplifications and zero log ratios to neutral regions in which no change occurred.

9.2.3 Recurrent HMM

As was stated in the introduction, we will use a recurrent hidden Markov Model (HMM) proposed by Shah *et al.* (2007) for the identification of extended chromosomal regions of altered copy numbers labeled as gain or loss [4]. The goal of this model is to construct features that distinguish the sporadic from the BRCA1 related group and subsequently to use them in a classifier (see Section 9.2.4). Because of the sensitivity of traditional HMMs to outliers being measurement noise, mislabeling and copy number polymorphisms in the normal human population, a robust HMM was first proposed by Shah *et al.* (2006) which handles outliers and integrates prior knowledge about copy number polymorphisms into the analysis [16]. To further reduce the influence of various sources of noise on the detection of recurrent copy number alterations, Shah *et al.* (2007) extended the robust HMM to a multiple sample version in which array CGH experiments from a cohort of individuals are used to borrow statistical strength across samples instead of modeling each sample individually [4]. This makes even copy number alterations in a small number of adjacent clones reliable when shared across many samples.

In this study, a recurrent HMM is constructed on a chromosomal basis separately for the group of sporadic and the group of BRCA1 mutated ovarian cancer. Both HMMs result in chromosomal regions with genetic alterations characterizing sporadic samples and samples with a BRCA1 mutation, respectively. A differential region is defined as a chromosomal region which is gained/lost in one group while not being gained/lost in the other group.

9.2.4 Kernel Methods and Least Squares Support Vector Machines

The differential regions that result from the recurrent HMM are used as features in a classifier for which we chose kernel methods. These methods are a group of algorithms that do not depend on the nature of the data because they represent data entities through a set of pairwise comparisons called the kernel matrix [17]. This matrix can be geometrically expressed as a transformation of each data point x to a high dimensional feature space with the mapping function $\Phi(x)$. By defining a kernel function $k(x_k, x_l)$ as the inner product $\langle \Phi(x_k), \Phi(x_l) \rangle$ of two data points x_k and x_l, an explicit representation of $\Phi(x)$ in the feature space is not needed anymore. Any symmetric, positive semidefinite function is a valid kernel function, resulting in many possible kernels, e.g. linear, polynomial and diffusion kernels. In this manuscript, a linear kernel function was used.

An example of a kernel algorithm for supervised classification is the Support Vector Machine (SVM) developed by Vapnik [18] and others. Contrary to most other classification methods and due to the way data is represented through kernels, SVMs can tackle high dimensional data (e.g. microarray data). The SVM forms a linear discriminant boundary in feature space with maximum distance between samples of the two considered classes. This corresponds to a non-linear discriminant function in the original input space. This kernel method also contains regularization which allows tackling the problem of overfitting. It has been shown that regularization seems to be very important when applying classification methods on high dimensional data [9]. A modified version of SVM, the Least Squares Support Vector Machine (LS-SVM), was developed by Suykens *et al.* (1999), (2002) [10]-[11]. On high dimensional data sets, this modified version is much faster for classification because a linear system instead of a quadratic programming problem needs to be solved.

9.2.5 Feature Selection

The choice of a feature selection technique is a widely discussed topic [19]. Lai *et al.* (2006) found that univariate gene selection, computationally simple and fast for high dimensional data, leads to good and stable performances across many cancer types and yields in many cases consistently better results than multivariate approaches [20]. Therefore, we will use a univariate method. Because no comparison of univariate gene selection techniques has been made across a sufficiently wide range of benchmark data sets and due to the dependency of the best performing technique on the data set used, Yang *et al.* (2005) proposed a method in which some of the arbitrariness with which univariate methods are chosen for high dimensional data is vanished [5]. This technique, called DEDS (Differential Expression via Distance Synthesis) is based on the integration of different test statistics via a distance synthesis scheme because features highly ranked simultaneously by multiple measures are more likely to be differential expressed than features highly ranked by a single measure. The statistical tests which were combined are ordinary fold changes, ordinary t-statistics, SAM-statistics and moderated t-statistics. The performance of DEDS is favorably comparable with the best individual statistic which is in practice often unknown and which depends on the data set used. Additionally, DEDS is not adversely affected by the worst performing statistic and achieves robustness properties which are lacked by the individual statistics. DEDS is available as a BioConductor package in R.

9.2.6 Proposed Methodology

Due to the limited number of samples, a leave-one-out (LOO) cross-validation strategy is applied. The 4 different steps that have to be accomplished in each LOO iteration are shown in Figure 9.1. After leaving out one sample, a recurrent HMM (see Sect. 9.2.3) is constructed in step 1 for both groups of sporadic and BRCA1 mutated ovarian cancer to determine the chromosomal regions with genetic alterations that characterize each group. Combining these regions results in the chromosomal regions that are differential between the remaining n-1 sporadic and BRCA1 mutated

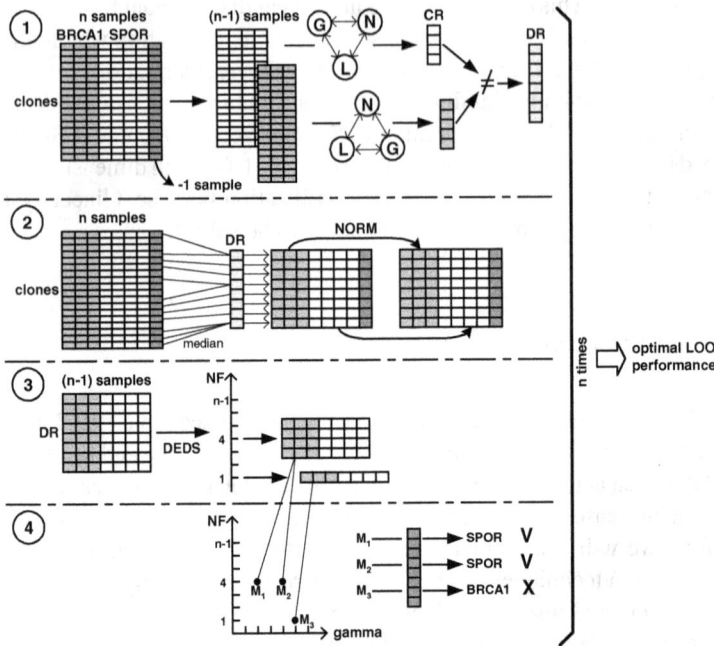

Fig. 9.1 Methodology consisting of 4 steps: step 1 - recurrent HMM; step 2 - conversion of clones to differential regions and normalization per sample; step 3 - feature selection using DEDS; step 4 - LS-SVM training and validation on left out sample (CR = Chromosomal Region; DR = Differential Region; NORM = Normalization; DEDS = Differential Expression via Distance Synthesis; NF = Number of Features)

samples. Because multiple clones can be located within each differential region, the clones need to be combined. This is done per sample in the second step by taking the median of the log ratios of the clones in each region. Afterwards, a standardization is performed per sample (i.e. meanshifting to 0 and autoscaling to 1) because the raw log ratios cannot be compared in absolute values between the samples. In step 3, DEDS determines which preprocessed log ratios, called features, best discriminate the n-1 samples (see Sect. 9.2.5). The number of included features is iteratively increased according to the obtained feature ranking without including more features than the number of samples on which the optimal number of features is determined [21]. This subset of features forms the input for classification in the last step (see Sect. 9.2.4). The LS-SVM contains a regularization parameter γ which, together with the number of features needs to be optimized. For all possible combinations of γ and number of features, an LS-SVM is built on the training set and validated on the left out sample. This is repeated n times such that each sample has been left out once. For the LS-SVM, a linear kernel function $k(x_k, x_l) = x_k^T x_l$ was chosen. An RBF kernel resulted in similar performances (data not shown).

9.2.7 Functional Annotation Analysis

To validate the selected chromosomal regions, gene set enrichment was performed as an indication for agreement with "known" biology. Two groups of gene sets as defined in the Molecular Signatures Database (MSigDB) were used: curated gene sets (i.e. sets of co-regulated genes from online pathway databases, publications in PubMed and knowledge of domain experts) and Gene Ontology (GO) gene sets (i.e. genes annotated by the same GO term) [13]. Using the HUGO gene nomenclature[1] [22], genes within the differential chromosomal regions were divided into 9 gene signatures, depending on the group (BRCA1 versus sporadic versus both) and CNV type (gain versus loss versus both). For each signature, the overlap was calculated between all gene sets and the signature and 5000 equally-sized signatures containing genes randomly selected from the genome. The corrected method of North *et al.* (2002) was used to calculate the empirical p-value for each gene set as $(r+1)/(n+1)$ with n the number of random signatures (i.e. 5000) and r the number of them with an equal or higher overlap with the gene set than obtained with the actual signature [23]. Only gene sets with r smaller than 10 (p-value < 0.002) were further investigated.

9.3 Results

Eight sporadic and five BRCA1 mutated ovarian cancer patients were included in this study and profiled using array CGH technology. Figure 9.2 gives an impression of array CGH data with which chromosomal regions that are different between 2 classes of samples can be identified. This figure shows an example of a recurrent amplification in BRCA1 patients which is not present in sporadic patients.

When applying the proposed methodology on this data set, CNVs in 11 chromosomal regions were sufficient to correctly classify 11 out of 13 samples. The LS-SVM had a LOO accuracy of 84.6%, a sensitivity of 100% (5/5) and a specificity of 75% (6/8).

Table 9.1 and Figure 9.3 show information on the 11 differential regions. Five regions are gained and 3 lost in BRCA1 mutated samples while the sporadic ovarian cancer patients are characterized by loss of 3 regions. A comparison of the 11 regions found in each of the 13 LOO iterations shows a limited variability in the selected regions. Table 9.1 also shows the number of LOO iterations in which each feature resulting from the complete data set is chosen which indicates stability of the 11 regions. The top 5 of features with the lowest p-value according to DEDS appeared in 8 to 11 of the 13 LOO iterations. Three less significantly features appeared in 4 LOO iterations.

Because we hypothesize that genes in the 11 chromosomal regions participate in processes that distinguish sporadic from hereditary ovarian cancer, a gene set

---

[1] http://www.genenames.org

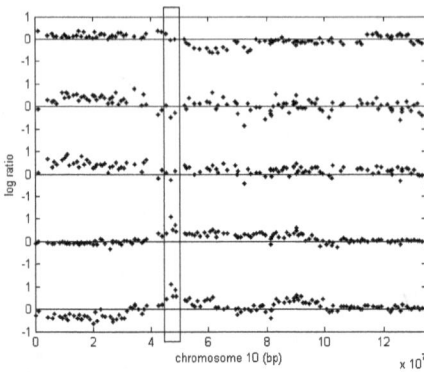

Fig. 9.2 Array CGH profile of chromosome 10 for 3 sporadic (top) and 2 BRCA1 mutated samples (bottom). The horizontal lines indicate the 0 log ratios for all samples. Both groups have a different profile within the first 3×10^7 base pairs and an amplification indicated with the vertical box occurs within the BRCA1 mutated samples around 5×10^7 base pairs.

Table 9.1 Chromosomal information on the 11 differential regions with the number of LOO iterations in which each of these regions was selected

Feature	Chromosome	Group	CNV type	Startbase	Stopbase	nb genes	nb LOO iter
1	13	BRCA1	loss	55423625	55550461	0	8
2	23	BRCA1	gain	3273880	7085387	5	11$^{\mu}$
3	12	BRCA1	gain	101502349	101656438	0	9
4	4	BRCA1	gain	10384154	19905375	22	11$^{\varsigma}$
5	4	sporadic	loss	4932958	8382645	24	10
6	3	BRCA1	gain	24167220	35751756	32	5
7	10	BRCA1	loss	4290650	17074128	66	7$^{\varsigma}$
8	16	sporadic	loss	56587489	67418517	81	4
9	19	BRCA1	loss	12159479	13216789	39	4
10	6	BRCA1	gain	24267702	29367215	86	4$^{\varsigma}$
11	16	sporadic	loss	70089429	75199166	36	6

$^{\mu}$Approximate correlation with LOO: region 10-50% smaller in 2 LOO runs
$^{\varsigma}$Approximate correlation with LOO: region 10-40% smaller in 1 LOO run

enrichment-based approach was followed (see Sect. 9.2.7). The most important gene sets enriched in the signatures are summarized below.

One of the components of the human SWI/SNF complex, regulating gene expression by remodeling nucleosomal structure in an ATP-dependent manner, is the gene BAF57 (a BRG1-associated factor). This gene mediates interaction with transcriptional activators or repressors and mutation of this gene has been found to be associated with a wide variety of tumours [24]. It is known that there is a direct interaction between BRG1- and hBRM-associated factors and the BRCA1 tumour suppressor protein. The human SWI/SNF complex affects cell growth and proliferation by interacting with tumour suppressor pathways and probably controlling them. Recent studies have shown the importance of complexes containing BAF57

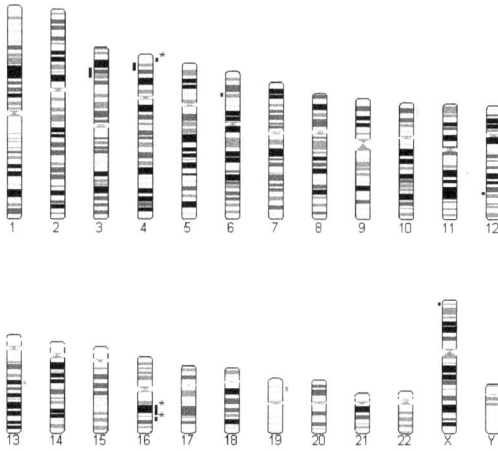

Fig. 9.3 BRCA1 - five gained regions (shown at the left of the chromosomes) and three lost regions (shown at the right in gray); sporadic - three lost regions (shown at the right indicated with the symbol *).

in transcriptional repression of tumour suppressor genes among which BRCA1. Wang and colleagues found 410 up-regulated and 469 down-regulated genes in cells with BAF57 re-expressed. Ten of the down-regulated genes (i.e. MED28, SUSD5, NCAPG, SLC4A7, MXRA5, MRS2, BST1, QDPR, LAP3, HS3ST1; p-value $< 2\mathrm{x}10^{-4}$) were found in four of the five regions gained in the BRCA1 mutated samples.

Another gene set consisting of 96 genes down-regulated at any time point (1-24 hours) following treatment of mammary carcinoma cells with exogenous human growth hormone (hGH) [25] was significantly overrepresented in the regions gained in BRCA1 ovarian cancer with an overlap of 7 genes (GPLD1, HIST1H2BK, HS3ST1, SLC4A7, SLC17A1; p-value $= 8\mathrm{x}10^{-4}$).

Many HOX genes, a subset of the homeobox genes, were recently found to be aberrantly expressed in a variety of cancers among which breast, kidney and skin suggesting that these HOX genes contribute to the progression of tumours. The homeobox HOXA5 encodes a transcriptional factor with an important role in embryogenesis, hematopoiesis and tumorigenesis. In human, it has been shown that HOXA5 mRNA levels are markedly reduced or even lost in more than 60% of breast cancer cell lines and primary breast carcinoma cells. This suggests that HOXA5 may act as a tumour suppressor gene in breast cells which makes loss of expression of this gene an important step in tumorigenesis [26]. Six genes, normally up-regulated in HOXA5-induced cells (with HOXA5 being a positive regulator), were found to be lost in BRCA1 ovarian cancer (ZNF44, DCLRE1C, ZNF136, KIN, JUNB, IER2; p-value $= 8\mathrm{x}10^{-4}$).

Tumour necrosis factor alpha (TNFα) is a proinflammatory cytokine with important roles in control of immune and inflammatory responses as well as cell cycle proliferation and apoptosis [27]. Of the genes up-regulated in TNFα-induced HeLa cells, four were found in 2 regions lost in BRCA1 ovarian cancer (IER2, PRDX2, JUNB, GDI2; p-value = 1.4×10^{-3}).

Three highly related Myb transcription factors (i.e. A-Myb, B-Myb and c-Myb) are expressed in vertebrates. The c-Myb gene, the proto-oncogene progenitor of the v-myb oncogene, is highly expressed in a.o. pancreatic, colon and breast tumours and his expression correlates with proliferation. A functional c-Myb protein is required for normal hematopoiesis. The A-Myb gene is expressed in a subset of the cells that expresses c-Myb [28]. Sporadic ovarian cancer is characterized by a loss of 9 genes activated by A-Myb or c-Myb genes (ATP6V0D1, MMP15, RRAD, S100P, E2F4, CTCF, PSMD7, CDH1, NFATC3; p-value = 6×10^{-4}).

9.4 Conclusion

In this manuscript, a new methodology is proposed in which copy number variations resulting from array CGH are transformed into features for classification purpose. This general method which is independent of cancer site allows to find a small set of chromosomal regions for distinguishing two classes of patients and to biologically validating them. It can also result in clinically relevant models based on a limited set of features. As increasing amounts of array CGH data become available, there is a need for algorithms to identify recurrent gains and losses based on statistically sound methods and to use them for classification. A large number of approaches for the analysis of array CGH data have already been proposed recently, ranging from mixture models and HMMs to wavelets and genetic algorithms [2]. However, most cancer studies that gather array CGH data only apply methods for exploratory analysis. Often a fixed threshold is used for defining gains and losses making these studies less robust against systematic changes in the baseline copy number measurements between samples [29]. A HMM on the contrary is a probabilistic method that can handle the uncertainty in the data in a formal way compared to deterministic algorithms. This makes the HMM more robust against outliers such as measurement noise and wrong recordings of locations of clones. Moreover, we used a special variant of HMM able to capture recurrent copy number alterations by coupling the HMMs of individual samples. This makes weak copy number alterations but shared across many samples reliable features. In our setup we used this property by first modeling the copy number variations in the group of sporadic and BRCA1 mutated patients separately. Subsequently, the alterations that were different between these two groups were used as features in an LS-SVM for classification. In our opinion this is one step further compared to many other studies that only perform an exploratory analysis.

The stability of the regions selected in each of the LOO iterations strengthens our confidence that the chromosomal regions found with our methodology are robust.

Two of the regions lacking genes with an annotated HUGO symbol seem uninteresting at first sight. However, recent research findings on 1% of the genome indicated that 93% of the bases are transcribed, increasing the importance of non-protein-coding RNA [30]. The remaining 9 regions were validated biologically using a gene set enrichment-based approach. Keep in mind that, because the number of features is minimized, one can expect that biological validation using pathways may fail because not all genes belonging to a certain pathway may be needed in a classification setting. In our subset the genes BAF57 and HOXA5 seemed to be correlated with hereditary ovarian cancer, whereas loss of the v-myb oncogene seemed more characteristic for the sporadic group.

Acknowledgements

AD is research assistant of the Fund for Scientific Research - Flanders (FWO-Vlaanderen). BDM is a full professor at the Katholieke Universiteit Leuven, Belgium. This work is partially supported by: **1.** Research Council KUL: GOA AM-BioRICS, CoE EF/05/007 SymBioSys, PROMETA, several PhD/postdoc & fellow grants. **2.** Flemish Government: **a.** FWO: PhD/postdoc grants, projects G.0241.04 (Functional Genomics), G.0499.04 (Statistics), G.0318.05 (subfunctionalization), G.0302.07 (SVM/Kernel), research communities (ICCoS, ANMMM, MLDM); **b.** IWT: PhD Grants, GBOU-McKnow-E (Knowledge management algorithms), GBOU-ANA (biosensors), TAD-BioScope-IT, Silicos; SBO-BioFrame, SBO-MoKa, TBM-Endometriosis. **3.** Belgian Federal Science Policy Office: IUAP P6/25 (BioMaG-Net, Bioinformatics and Modeling: from Genomes to Networks, 2007-2011). **4.** EU-RTD: ERNSI: European Research Network on System Identification; FP6-NoE Biopattern; FP6-IP e-Tumours, FP6-MC-EST Bioptrain, FP6-STREP Strokemap.

References

1. Pinkel, D., Albertson, D.G.: Array comparative genomic hybridization and its applications in cancer. Nature Genetics **37**(Suppl.) (2005) 11–17
2. Lai, W.R., Johnson, M.D. *et al.*: Comparative analysis of algorithms for identifying amplifications and deletions in array CGH data. Bioinformatics **21**(19) (2005) 3763–3770
3. Guha, S., Li, Y. *et al.*: Bayesian hidden markov modeling of array CGH data. Harvard University Biostatistics Working Paper Series, Working paper 24 (October 2006)
4. Shah, S., Lam, W.L. *et al.*: Modeling recurrent DNA copy number alterations in array CGH data. Bioinformatics **23** (2007) i450–i458
5. Yang, Y.H., Xiao, Y. *et al.*: Identifying differentially expressed genes from microarray experiments via statistic synthesis. Bioinformatics **21**(7) (2005) 1084–1093
6. Cristianini, N., Shawe-Taylor, J.: An introduction to Support Vector Machines and other kernel-based learning methods. Cambridge University Press, Cambridge (2000)
7. Shawe-Taylor, J., Cristianini, N.: Kernel methods for pattern analysis. Cambridge University Press, Cambridge (2004)

8. Bhaskar, H., Hoyle, D.C. *et al.*: Machine learning in bioinformatics: A brief survey and recommendations for practitioners. Computers in Biology and Medicine **36**(10) (2006) 1104–1125

9. Pochet, N., De Smet, F. *et al.*: Systematic benchmarking of microarray data classification: assessing the role of nonlinearity and dimensionality reduction. Bioinformatics **20** (2004) 3185–3195

10. Suykens, J., Vandewalle, J.: Least Squares Support Vector Machine classifiers. Neural Processing Letters **9** (1999) 293–300

11. Suykens, J.A.K., Van Gestel, T. *et al.*: Least Squares Support Vector Machines. World Scientific, Singapore (2002)

12. Gajewski, W., Legare, R.D.: Ovarian cancer. Surgical Oncology Clinics of North America **7** (1998) 317–333

13. Subramanian, A., Tamayo, P. *et al.*: Gene set enrichment analysis: A knowledge-based approach for interpreting genome-wide expression profiles. Proceedings of the National Academy of Sciences **102**(43) (2005) 15545–15550

14. Burke, W., Daly, M. *et al.*: Recommendations for follow-up care of individuals with an inherited predisposition to cancer. II. BRCA1 and BRCA2. Cancer Genetics Studies Consortium. Journal of the American Medical Association **277** (1997) 997–1003

15. Starita, L.M., Parvin, J.D.: The multiple nuclear functions of BRCA1: trancription, ubiquitination and DNA repair. Current Opinion in Cell Biology **15**(3) (2003) 345–350.

16. Shah, S., Xuan, X. *et al.*: Integrating copy number polymorphisms into array CGH analysis using a robust HMM. Bioinformatics **22**(14) (2006) e431–e439

17. Schölkopf, B., Tsuda, K. *et al.*: Kernel methods in computational biology. MIT Press, United States (2004)

18. Vapnik V.: Statistical Learning Theory. Wiley, New York (1998)

19. Saeys, Y., Inza, I. *et al.*: A review of feature selection techniques in bioinformatics. Bioinformatics **23**(19) (2007) 2507–2517

20. Lai, C., Reinders, M.J.T. *et al.*: A comparison of univariate and multivariate gene selection techniques for classification of cancer datasets. BMC Bioinformatics **7** (2006) 235–244

21. Li, W., Yang, Y.: How many genes are needed for a discriminant microarray data analysis. In Methods of Microarray Data Analysis, eds = Lin, S.M. and Johnson, K.F., Kluwer Academic (2002) 137–150

22. Wain, H.M., Bruford, E.A. *et al.*: Guidelines for human gene nomenclature. Genomics **79**(4) (2002) 464–470

23. North, B.V., Curtis, D. *et al.*: A note on the calculation of empirical p values from Monte Carlo procedures. American Journal of Human Genetics **71** (2002) 439–441

24. Wang, L., Baiocchi, R.A. *et al.*: The BRG1- and hBRM-associated factor BAF57 induces apoptosis by stimulating expression of the cylindromatosis tumor suppressor gene. Molecular and Cellular Biology **25**(18) (2005) 7953–7965

25. Xu, X.Q., Emerald, S. *et al.*: Gene expression profiling to identify oncogenic determinants of autocrine human growth hormone in human mammary carcinoma. The Journal of Biological Chemistry **280**(25) (2005) 23987–24003

26. Chen, H., Rubin, E. *et al.*: Identification of transcriptional targets of HOXA5. The Journal of Biological Chemistry **280**(19) (2005) 19373–19380

27. Zhou, A., Scoggin, S. *et al.*: Identification of NF-κB-regulated genes induced by TNFα utilizing expression profiling and RNA interference. Oncogene **22** (2003) 2054–2064

28. Lei, W., Rushton, J.J. *et al.*: Positive and negative determinants of target gene specificity in Myb transcription factors. The Journal of Biological Chemistry **279**(28) (2004) 29519–29527

29. Klijn, C., Holstege, H. *et al.*: Identification of cancer genes using a statistical framework for multiexperiment analysis of nondiscretized array CGH data. Nucleic Acids Research **36**(2) e13

30. The ENCODE Project Consortium: Identification and analysis of functional elements in 1% of the human genome by the ENCODE pilot project. Nature **447** 799–816

Author Index

www.ingramcontent.com/pod-product-compliance
Lightning Source LLC
Chambersburg PA
CBHW061333220326
41599CB00026B/5157